Smart Thermodynamics

Smart Thermodynamics

Lokesh Pandey

ARCLER
PRESS

www.arclerpress.com

Smart Thermodynamics

Lokesh Pandey

Arcler Press
224 Shoreacres Road
Burlington, ON L7L 2H2
Canada
www.arclerpress.com
Email: orders@arclereducation.com

ISBN: 978-1-77407-613-2 (Hardcover)

ABOUT THE AUTHOR

Lokesh Pandey is currently pursuing his PhD in Mechanical Engineeing from Uttarakhand Technical University, India, where he also completed his M.Tech in Thermal Engineeing. He has more than 3 years teaching experience as a Faculty for Mechanical Engineering. He has published articles in reputed Journals and has also chaired many conferences and workshops.

TABLE OF CONTENTS

LIST OF FIGURES

LIST OF TABLES

LIST OF ABBREVIATIONS

°F	Fahrenheit scale
AARD	average absolute relative deviation
Ar	Argon
A^R	Helmholtz energy
BTU	British thermal unit
C	carbon
Ca^{2+}	calcium
CAD	computer aided drafting
CALPHAD	calculation of phase diagrams
Cl^-	chloride
COP	coefficient of performance
COPHP	coefficient of performance for heat pump
COPR	coefficient of performance for refrigerators
Cr	chromium
DSC	differential scanning calorimetry
DTA	differential thermal analysis
EOS	equation of state
GNI	gross national income
H^+	hydrogen
HCl	hydrogen chloride
HCO_3^-	hydrogen carbonate
He	helium
HPO_4^{2-}	hydrogen phosphate
ICME	integrated computational materials engineering
ITC	isothermal titration calorimetry
K	Kelvin scale
K^+	potassium

LPG	liquefied petroleum gas
Mg^{2+}	magnesium
MGI	materials genome initiative
MPa	megapascals
MTC	Mattedi-Tavares-Castier
Na^{+}	sodium
NaCl	sodium chloride
NG	natural gas
NRTL	non-random two-liquid
P	pressure
p-h	pressure-enthalpy
PMM	perpetual motion machine
SAFT	statistical associating fluid theory
SER	standard element reference
SGTE	scientific group thermo data Europe
SHS	self-extending high-temperature synthesis
$SOCl_2$	thionyl chloride
STP	standard thermodynamic properties
T	temperature
TCPC	three-characteristic-parameter correlation
TD	thermodynamic databases
Ti	Titanium
TM	thermodynamic modeling
T-MATS	Toolbox for the Modeling and Analysis of Thermodynamic Systems
T-s	temperature-entropy
UNIQUAC	universal quasi chemical
V	volume
VSST	vehicle systems safety technologies

PREFACE

Thermodynamics is an interesting subject that deals with heat and temperature and their relation to energy, work, radiation, and properties of matter. It is a discipline which has an exceptionally wide range of applicability and is the most ubiquitous subfield of Physics which is studied by Engineers, Chemists, and Material Scientists as well.

Science is mainly concerned with explaining different phenomena around the world. Thermodynamics alone explains the occurrence of more things around us than all other physical theories which are studied in the Physics curriculum.

Thermodynamics explains concepts like why heat flows from hot to cold bodies, why the Sun appears yellow, why the air is colder and thinner in the mountains, why temperature below -273 centigrade cannot be measured and much more.

Thermodynamics is of practical importance, because it explains the processes that transforms heat into other forms of energy like electrical potential and mechanical work. The laws of thermodynamics govern behavior of chemical reactions. Hence, it is applied in diverse areas ranging from microscopic organisms to power generation systems, transportation vehicles, pharmaceuticals power plants and common household appliances.

Lord Kelvin, an Irish physicist was the first to come up with a definition for thermodynamics in 1854, when he stated "Thermo-dynamics is the subject of the relation of heat to forces acting between contiguous parts of bodies, and the relation of heat to electrical agency."

Thermodynamics was a discipline which was born out of the necessity to increase the efficiency of steam engines. The French physicist Nicolas Léonard Sadi Carnot discovered the Carnot cycle and was successful in increasing the efficiency of steam engines.

Application of thermodynamics to mechanical heat engines was also extended to the study of chemical compounds and reaction which helped in the expansion of knowledge of this field. It helped in understanding the role of entropy in the process of chemical reactions. Gradually other formulation of thermodynamics such as computational thermodynamics emerged.

The foundation of thermodynamics is a set of four laws, namely the zeroth law, first law, second law, and third law. These universally valid when it is applied to systems which operate within the limitations set by each law. These laws govern the chemical processes in all biological systems and important unifying principles in biology as well.

This book covers all the important concepts in thermodynamics. The book starts with the basic concepts and moves on to more complex topics such as entropy, reversibility, and irreversibility in thermodynamics, stability of thermodynamic systems, computational thermodynamics and thermodynamic modeling and simulation.

This book is written in a simple manner which is easy for the readers to understand and grasp the various concepts of thermodynamics. This discipline is a critical area for students of science and technology. Besides, the analogous concepts are also useful for computer science and social science students.

CHAPTER 1

Introduction to Thermodynamics

CONTENTS

Thermodynamics is an important branch of science which is concerned with energy and work of a system. It specifically explains how thermal energy is converted into other forms of energy and the impact of this on matter.

This chapter first explains the history of thermodynamics and highlights the important events in the course of evolution of the subject. There are various branches and sub-branches of this discipline. The chapter focuses on four important branches of thermodynamics.

The chapter then describes the most important concepts of thermodynamics namely work and heat and explains various aspects like specific heat and heat transfer. The chapter then explains other important concepts like temperature and entropy.

Thermodynamic systems are macroscopic regions of universe which is being studied. The chapter explains the properties of system, state, and its processes. The chapter then explains the properties of a pure system. Finally, the chapter describes Rankine cycle which is also called steam cycle and explains the meaning of Carnot cycle.

1.1. INTRODUCTION

Thermodynamics is a branch of physics which is mainly concerned with the heat and temperature. It deals with the relation of these two quantities with energy, radiation, work, and properties of a matter. It specifically describes the manner in which thermal energy is converted into other forms of energy and the impact it has on matter.

Thermal energy is the energy which a matter or system possess due to its temperature. The molecules move or keep vibrating which causes the temperature. Thermodynamics enables the measurement of this energy which is a complex science.

According to David McKee, a professor of physics at Missouri Southern State University. Thermodynamics involves measuring this energy, which can be "exceedingly complicated. The systems that we study in thermodynamics consist of very large numbers of atoms or molecules interacting in complicated ways. But, if these systems meet the right criteria, which we call equilibrium, they can be described with a very small number of measurements or numbers. Often this is idealized as the mass of the system, the pressure of the system, and the volume of the system, or someother equivalent set of numbers. Three numbers describe 1026 or 1030 nominal independent variables."

The behavior of heat and temperature is explained by the laws of thermodynamics. These laws are first law, second law, and third law. The zeroth law of thermodynamics which was conceived later, is considered in nature and hence precedes the first law.

These laws convey a quantitative description using measurable macroscopic physical quantities. It can also be elucidated in terms of microscopic constituents. This aspect is also dealt in statistical mechanics. Thermodynamics comprises of wide range of topics in science and engineering. In particular, it involves chemical engineering, physical chemistry, and mechanical engineering, as well as complex fields such as meteorology.

The method primarily used in thermodynamics is experimental observation which are formalized into the basic laws of the subject.

Thermodynamics is also concerned with the mathematical modeling of the real world as in the case of other sciences. To ensure the consistency of mathematical deductions, certain basic concepts and precise definitions have to be established. These important concepts are discussed in this chapter.

1.2. HISTORY OF THERMODYNAMICS

Thermodynamics is a wide branch of science and has a long historical development from the ancient period till the 20th century. The invention of the thermometer was a landmark development which facilitated the first precise speculations on heat.

In the 1600s, the basic physical notions of heat and temperature were established. During that time, the scientists associated heat with the motion of microscopic constituents of matter. However, in the 1700s the notion that heat is separate fluid-like state was conceived.

In the 1840s, James Joule and others conducted experiments which raised doubts about the state of heat. Finally, in the 1850s heat was accepted as a form of energy. This was a very important development which then paved way for invention of steam engine. In 1824, Said Carnot captured some ideas of thermodynamics when he explained the efficiency of idealized engine.

There was an absence of theories about the nature of heat till a considerable time. The majority of experts in science, viewed heat as a substance. The Kinetic Theory was rejected did not receive attention from the scientists. The Calorific Theory dominated till the 1850s and was used to explain natural phenomena like gas laws and heat transfer. This continued till the time the Principle of Conservation of Energy was introduced. Around the 1850s Rudolf Clausius and William Thomson (Kelvin) stated the First Law and Second Law of Thermodynamics.

The Second Law was formulated on the basis that heat does not flow automatically from colder to hotter body. The other formulations were derived from this. Kevin was able to understand some general implications of the law. In 1738, Daniel Bernoulli discussed in detail the idea that gases comprise of molecules, but this idea was not supported much. It was revived by Clausius in 1857.

In 1860 James Clerk Maxwell derived the expected distribution of molecular speeds in gas from the mechanics of individual molecular collisions. During the years that followed, the kinetic theory of gases developed rapidly which facilitated the computation of several microscopic properties of gases in equilibrium. The Kinetic Theory of Gases provided the impetus to scientists to introduce the concept of entropy. This facilitated the formulation of laws of thermodynamics in a perfect mathematical form. This paved the way for a new branch of physics called statistical thermodynamics.

In 1872 Ludwig Boltzmann came up with an equation which gave a detailed description of time development of a gas, both in equilibrium and non-equilibrium state. In the 1860s Clausius had discussed about entropy as a ratio of heat to temperature and came up with the Second Law in terms of the increase of this quantity. He then demonstrated that his equation implied the so-called H Theorem according to which, a quantity which equals entropy in equilibrium must always increase in time.

It appeared that the Boltzmann was successful in his attempt to prove the Second Law. However, it was realized that since it was assumed that molecular collisions were reversible, his derivation could be run in reverse. Hence the explanation of the Second Law could be questioned.

It was realized later that while Boltzmann's original equation implicitly assumed that molecules were not related to each other before collision, but this assumption was not held post-collision. This introduced a fundamental asymmetry in time. It seems that early in 1870s Maxwell and Kelvin understood that the Second Law could not be formally derived from microscopic physics but must somehow be a consequence of human inability to track large numbers of molecules.

There were objections concerning reversibility. Consequently, Boltzmann realized around 1876 that gas can be of many more states that seem random than seem orderly.

As a result, he argued that entropy must be proportional to the logarithm of the number of possible states of a system. He then formulated several ideas about ergodicity. In the beginning of 1900, Willard Gibbs put the statistical mechanics of particles in a general context.

Gibbs came up with the concept of an ensemble. It refers to a collection of several possible states of a system each of which has been assigned a certain possibility. His explanation was that if the time evolution of a single state goes through all other states in the ensemble, then when the average is calculated over a long period of time, a single state would exhibit the typical behavior of the ensemble.

Gibbs also presented qualitative arguments about entropy. He asserted that when measured in a coarse-grained way in which no distinction is made between the nearby states, the entropy would increase.

In the early 1900s, quantum theory largely dominated the field and not much fundamental work was done for the development of thermodynamics. However, by 1930s the Second Law began to be recognized as a principle of physics which largely remained unquestioned.

Though the ergodic theory was not focused in physics, it became an active area in pure mathematics. During the period 1920s to the 1960s, the several properties of ergodicity were established for many of the simple systems.

In 1950s the electronic computers came into existence. Enrico Fermi and others started their investigation about the ergodic properties of nonlinear system of springs. However, they majorly focused on recurrence phenomena related to solitons and did not consider the general questions related to Second Law.

During the 1950s, it began to be believed that entropy is simply the negative of the information quantity. The latter was introduced by Claude Shannon in 1940s. John von Neumann made statements which led to the belief that any computational process is accompanied by an increase in entropy. However, this notion was rejected, specially due to the works of Charles Bennett. His works laid the foundation for relating computational and thermodynamic ideas.

In 1960s, the first simulations of hard sphere gases were led to concentrate on the specific phenomenon of long-time tails and once again the general questions related to Second Law was not considered. Computer experiments gradually became oriented towards ordinary differential equations and strange attractors and not on systems with several components. The Second Law was yet not focused on.

In the beginning of the 20th century, the concept of thermodynamic potentials and absolute temperature scale were introduced post which the Third Law of Dynamics was discovered. During the time the scientific issue of thermal radiation was solved.

As discussed, thermodynamics has a long history. However, its development was not as crisp as mechanics which progressed due to the contribution of Newton. Its development is characterized by false steps, debates, and errors which continues even today. He milestones of its development are discussed below:

First century AD: Many thermal engines were documented in the works of Hero of Alexandria;

- 1593: Galileo invented the water thermometer;
- 1650: The first vacuum pump is designed and built by Otto von Guericke;
- 1662: The law for Isothermal ideal gases is developed by Robert Boyle;
- 1679: The steam digester which was the forerunner to the steam engine was developed by Denis Papin.;
- 1698: Thomas Savery an English inventor and engineer patents an early steam engine which is the first commercially used device for this purpose;
- 1710: Thomas Newcomen invented the first practical fuel-burning steam engine;
- 1760s: Joseph Black a Scottish physicist and chemist developed a calorimetry;
- 1780s: James Watt made further improvements to the steam engine;
- 1798: Benjamin Thompson (Count Rumford) cannon boring experiments and started contemplating about the mechanical equivalent of heat from it;

- 1824: Nicolas Leonard Sadi Carnot elaborated about idealized heat engines;
- 1840s: Julius Robert von Mayer and James Prescott Joule relate heat and work;
- 1840: Germain Henri Hess came up with first law of thermodynamics involving work-free chemical reactions;
- 1847: Hermann von Helmholtz conceived the theory of energy conservation and published it;
- 1848: William Thomson (Lord Kelvin) came up with an absolute zero of temperature;
- 1850: Rudolf Julius Emanuel Clausius, one of the central founders of thermodynamics, formalized the second law of thermodynamics;
- 1865: The concept of entropy came into existence due to the contribution of Clausius;
- 1871: James Clerk Maxwell postulated the Maxwell relations;
- 1870s: Josiah Willard Gibbs further validated mathematical thermodynamics. His model provides one single differential equation of all thermodynamic properties;
- 1870s: Statistical thermodynamics emerged due to the contribution of Maxwell and Ludwig Boltzmann; and
- 1889: Gibbs provided underlying foundations for classical and statistical thermodynamics by developing statistical mechanics.

Much development continued in the twentieth century, with pioneering work by Nobel laureates. Some of them were Jacobus Henricus van't Hoff (1901), Heike Kamerlingh Onnes (1913), Walther Nernst (1920), Albert Einstein (1921), Percy Bridgman (1946), and Kenneth Wilson (1982).

Sir Issac Newton's work was also concerned with the field of thermodynamics. He concentrated on energy; however, they are more related to the mechanical energy. He did not consider the notion that thermal energy exists, and it could be equated to mechanical energy.

Note, however, that temperature was known to Newton, as was Boyle's law. However, when he tried to apply his theories to problems of thermodynamics, such as calculation of the speed of sound in air, they notably failed. The reason for the failure required consideration of the yet-to-be-developed second law of thermodynamics.

1.3. BRANCHES OF THERMODYNAMICS

The roots of thermodynamics can be traced in thermochemistry, affinity-chemistry, and thermoelectricity. The foundation of this discipline was laid down by the works of Said Carnot, Rudolf Clausius, William Thomson, and William Rankine. Various branches and sub-branches of thermodynamics.

According to Elias Gyftopoulos and Gian Beretta, the subject of physics, in fact, science in general, is concerned with describing all aspects of all phenomenon which pertains to the perceivable universe. It is like a huge tree with several branches such as gravitation, mechanics, electromagnetism, and chemistry each of which focuses on a particular class of phenomenon. Thermodynamics is not a branch; it pervades the entire tree.

In the light of the above, the vastness of the discipline of thermodynamics can be perceived. In this chapter, the four main branches of thermodynamics are discussed.

1.3.1. Classical Thermodynamics

This branch is mainly concerned with the description of the states of thermodynamics systems which are near equilibrium It uses macroscopic and measurable properties of the systems for this purpose. It explains the exchange of work and heat on the basis of the laws of thermodynamics.

The qualifier classical indicates the fact it is the first level understanding of this subject. A microscopic interpretation of these concepts is dealt with by statistical mechanics which developed later.

In the history of thermodynamics, classical thermodynamics refers to the thermodynamics which developed before atomic structure was discerned. It is generally accepted that 1890s marked the beginning of classical period.

The terms seem to have an elusive coinage history. Georg Helm is believed to have written the first section which was devoted to a discussion on the use of the term classical. He included a section on classical thermodynamics in his 1898 *Historical Development of Energetics* in which he argued that the thermodynamics has reached as well-established state and hence can be referred to as classical. Therefore, the term classical thermodynamics is appropriate.

According to the American physicist Edgar Buckingham, it was Pierre Duhem, who introduced the term classical thermodynamics. Starting in 1950s, various books and textbooks were written using this title and which continued till the 1970s.

British physicist Brian Pippard published *Elements of Classical Thermodynamics* in 1957 which defined classical thermodynamics as "the method of approach that takes no account of the atomic constitution of matter [and] makes no attempt to provide a mechanistic explanation of why a given substance has the properties observed experimentally."

As per this definition, classical thermodynamics can be dated to the publication up until the year 1889 or the years before the thermodynamic publications of German physicist Max Planck and his theory of energy quanta in black bodies.

1.3.2. Statistical Thermodynamics

The origin of statistical mechanics also referred to as statistical thermodynamics can be traced in the late 19th and early 20th century when the atomic and molecular theories were developed. It provided an interpretation of the microscopic interactions between individual particles or quantum-mechanical states and supplemented the classical thermodynamics.

The subject is mainly concerned with the study of the average energy behavior of large groups of individual particles. It uses statistical methods to study random motion of molecules.

Statistical thermodynamics related the microscopic properties of individual atoms and molecules to the macroscopic properties. Thereby, at microscopic level, it explains classical thermodynamics as an outcome of statistics, quantum theory and classical mechanics.

The term "statistical thermodynamics" was first used in the 1930s when it was first loosely defined as statistical mechanics and started emphasizing on thermodynamics.

In the recent times, many authors such as social physicists have used statistical thermodynamics implicitly. For example, in human particle logic, as a means of explaining or arguing mass human system behavior.

1.3.3. Chemical Thermodynamics

Chemical thermodynamics is concerned with the study of interrelation of energy with physical change of state or chemical reactions based on the laws of thermodynamics. In other words, it is the energetic and entropic study of physical processes or chemical reactions. The fundamental aspects of the study are the first and the last stage of the process or reactions.

When energy and entropy are known, for all values of pressure, volume, and temperature, then a complete thermodynamic examination is possible. In this sense, chemical thermodynamics can be described as the study of entropy, energy, and equilibrium in chemical systems.

Chemical thermodynamics is considered as the most intellectually stimulating discipline in science. It is considered as the feeder course for chemical engineering.

The foundation of chemical thermodynamics was laid by the American engineer Willard Gibbs. Its roots can be traced to the eighteenth-century ideas of chemical affinity and force. These concepts transformed the conceptions of work and energy.

August Horstman, a German physical chemist, began to study the works of Rudolph Clausius with the aim of applying them to the calculation of equilibriums in chemical systems. In 1869, he published a paper "Dampfspannung und Verdampfungswärme des Salmiaks" (Vapor tension and heat of vaporization of the salmiac) in which for the first time, entropy was applied to chemical problems. In October 1873, he announced that maximum entropy was the condition for chemical equilibrium.

Another physicist Williard Gibs noted that the condition for thermodynamic equilibrium in a chemical system at constant temperature and pressure is that a certain function should be minimum. This function is now known as thermodynamic potential. Gibbs further elaborated on the subject of equilibriums and presented a comprehensive and rigorous treatise on the subject.

Horstmann made notable contribution to this field. There were two other physicists who were noted for their work and considered as central founders. One was Gibbs due to his publication *On the Equilibrium of Heterogeneous Substances,* in 1876. The other was German physicist and physician Hermann von Helmholtz, who mentored Horstmann. In 1882 he published *The Thermodynamics of Chemical Processes* (*Die Thermodynamik Chemischer Vorgänge*).

However, the emergence of this discipline is attributed to Gibbs's work in 1876. According to biographer Lynde Wheeler, Gibbs came up with the prime or fundamental equation

$$d\varepsilon = td\eta - pdv + \mu_1 dm_1 + \mu_2 m_2 \ldots\ldots + \mu_d m_n$$

which is represented in modern notation as:

$$dU = TdS - PdV + \sum_{i=1}^{n} \mu_i dm_i$$

This is the fundamental equation of Gibbs' *Equilibrium* treatise and marks the emergence of theory of chemical thermodynamics.

In 1913, James Partington, an English Chemist, published his *Textbook of Thermodynamics with Special Reference to Chemistry*. In 1924, he published an updated second edition of the same which was *Chemical Thermodynamics.* American physical chemists Gilbert Lewis and Merle Randall, are considered the founders of modern Chemical Thermodynamics. In1923, they published *Thermodynamics and the Free Energy of Chemical Substances*, in which for the first time the principles of Gibbs were applied specifically to chemical processes. English physical chemist Guggenheims also noted for his contribution. In 1933 he published *Modern Thermodynamics by the Methods of Willard Gibbs.* The collective contribution of these three chemists unified the science of chemical thermodynamics into a coherent whole.

1.3.4. Equilibrium Thermodynamics

Equilibrium thermodynamics is mainly concerned with transfers of matter and energy in systems or bodies that, by agencies in their surroundings, can change their thermodynamic equilibrium to another. In other words, it is concerned with the study of systems which are at equilibrium, approaching equilibrium or near equilibrium. 'Thermodynamic equilibrium' is a term used to indicate a state of balance. In this state, all macroscopic flows are zero. In this state the intensive properties of system or bodies become homogeneous and their pressures become perpendicular to their boundary.

Equilibrium thermodynamics is often used to distinguish between assumed continuous equilibrium states versus non-equilibrium thermodynamics. An example of the former is studies of rock formation and an example of latter is Bernard's cell formation. The "local equilibrium" criteria are used in the justification of use of equilibrium thermodynamic type equations.

Unbalanced potentials or driving forces between macroscopically distinct parts are not present in equilibrium. One of the main objectives of thermodynamics is to calculate the final equilibrium state of the system after a thermodynamic process has influenced its surroundings and walls from a given well-defined equilibrium state, surroundings, and constitutive walls.

Non-equilibrium thermodynamics is mainly concerned with systems which are not in thermodynamic equilibrium. Majority of the systems that occur in nature is not in thermodynamics equilibrium. They are not fixed but on a continuous or discontinuous basis subject to flux of energy and matter which takes place to and from systems.There are more general concepts in the thermodynamic study of non-equilibrium systems than involved in equilibrium thermodynamics. There are many natural systems which are not considered for macroscopic thermodynamic methods which are currently practiced.

1.4. FUNDAMENTAL CONCEPTS: WORK AND HEAT

It is important to understand the definitions of heat and work. Distinguishing between the two helps in making a correct analysis of several thermodynamic problems.

1.4.1. Definition of Work

Work is defined as a force F. It acts through a displacement x which is in the direction of the force.

$$W = \int_1^2 F dx$$

In thermodynamics, work is said to be done by a system if raising of a weight is the only effect on the surrounding. Surrounding refers to everything which is external to a system. When a system does a work, it is considered positive and when work is done to a system, it is considered negative.

Let us consider the example of a work crossing the boundary of the system. A battery and motor are an example of a system. The motor is driven by a fan which is outside the system. Here the work can be said to be crossing the system. If the fan is replaced and instead a pulley and a weight arrangement is used, the only effect occurring in the surroundings is the increase in weight.

In SI units, the unit for work is called the joule (J).

1J = 1 N•m

Numerical Example: The gas is a cylinder is a system. A piston is fitted to a cylinder in which a small weight is placed. The initial volume of gas and initial pressure are 0.04 m^3 and 200 kPa, respectively. A Bunsen is placed under the cylinder and the volume of gas increased to 0.1 m^3. The pressure remains unchanged. What is the work done by the system during the process?

$$W = \int_{1}^{2} PdV$$

Since the pressure is constant,

$$W = P\int_{1}^{2} dV = P\left(V_2 - V_1\right)$$

$$W = 200\text{kpa} \times (0.1 - 0.4)\text{m}^3 = 12\text{kj}$$

1.4.2. Definition of Heat

Heat is a form of energy. Heat is transferred from one system at a given temperature to another system or surroundings which is at a lower temperature. This occurs when there is a temperature difference between the two systems. Hence, heat crosses boundaries in this process.

A body does not contain heat. It is identified during the process it crosses the boundary. Therefore, heat is a transient and lasts only for a short time.

Let us consider the example of two systems, one of which is a hot block and the other is a cold water in a beaker. Both systems do not contain heat initially, both contain energy.

If the copper block is placed in the water, a thermal communication is established between the two. Transfer of heat takes place from copper to water till the time equilibrium of temperature is achieved.

Once it is achieved, heat transfer no longer takes place as the temperature is the same. The identification of heat occurs at the boundaries.

When heat is transferred to a system, it is considered positive, when it is transferred from a system it is considered negative. Heat is represented by the symbol Q.

Similar to work, heat is a form of energy transfer. It can occur both to and from the system. The unit used to denote heat and work are the same. In International System joule is the unit for heat.

1.4.3. Difference between Heat and Work

Both heat and work are ways by which energy is transferred from one system to another. As mentioned, this distinction is important to understand in the field of thermodynamics. The main difference between the two is that while heat is the transfer of thermal energy between the systems, work is the transfer of mechanical energy from one system to another.

Heat is the microscopic motion and work is macroscopic motion and their distinction is significant to understand how thermodynamic process work. Heat can be transferred to and from work. The first and second law of thermodynamics have discussed these two concepts. These are discussed in detail in the next chapterAccording to the first law of thermodynamics both heat and work contribute to the total internal energy of a system. The second law, however, limits the amount of heat that can be converted into work. Though this law allows work to be fully converted into heat, it does not allow heat to be fully transformed into work. This is because in the second case, the laws of entropy will be violated. The Carnot efficiency describes the maximum amount of work that can be attained from heat. The Carnot cycle is described in detail in chapter 5.

The other difference is that heat is an energy which is associated with random motion of particles. On the other hand, work is the energy of ordered motion in the same direction. Heat can therefore be considered as "low-quality" energy and work can be considered as "high-quality" energy. This notion aligns with the entropy statement of the Second Law.

Both heat and work possess distinct properties. They are quite different in the way they affect a system. The following list compares their properties:

Table 1.1. Comparison of Properties of Heat and Work

	Heat (Q)	**Work (W)**
Interaction	Thermal	Mechanical
Require-ment	Temperature Difference	Force and Displacement
Process	Microscopic collisions occur	Macroscopic pushes and pulls occur
Positive value	This is represented by $Q > 0$. This occurs when the environment is at a higher temperature compared to the system. The energy gets transferred into system.	This is represented by $W > 0$. Positive value occurs when a gas is compressed. Energy is gets transferred into system.
Negative value	This is represented by $Q < 0$. Negative value is when the system is at a higher temperature than the environment. Energy gets transferred out of system.	This is represented by $W < 0$. This occurs when a gas expands. Energy gets transferred out of system.
Equilibrium	A system is said to be in thermal equilibrium when its temperature and that of the environment is the same.	A system is said to be in mechanical equilibrium when there is an absence of net force or torque on it.

Source: Table by Encyclopedia.

1.4.4. Specific Heat

Specific heat refers to the amount of heat which is required to increase the temperature of a given mass of substance by a certain amount. It is also referred to as specific heat capacity. The convention unit of specific heat is calories per gram per kelvin. Calorie refers to the amount of heat energy used to increase the temperature by 1°C of 1 gram of water which is at 4°C.

In the case of metals, its specific heat is determined by the number of atoms in the sample. The mass does not influence the specific heat. For example, the atoms are more densely packed in aluminum than lead, hence one kilogram of aluminum can absorb 700 % heat of what one kilogram of lead can absorb. Lead atoms have a slightly higher heat absorption capacity compared to aluminum, 8 percent to be specific. The heat absorption capacity of a given mass of water is nearly five times higher when compared to the same mass of aluminum.

Measuring the specific heat of gas is complex. It can be measured at either constant pressure or constant volume and values vary in both cases.

1.4.5. Heat Transfer

As mentioned, the transfer of heat from one body to another or between a body and the environment takes place in three ways conduction, convection, and radiation. Conduction refers to the transfer of energy through a solid medium. In order for this to happen, direct contact between these two bodies is necessary. When in contact, the molecules transfer their energy across the interface.

Convection refers to the heat transfer to or from a fluid medium. The molecules present in a liquid or gas, when in contact with a solid body transmits heat to a body or absorbs from it. Each molecule then moves away, and other molecules take their place and the entire process is repeated.

The efficiency of heat transfer can be improved when the surface area to be heated or cooled is increased. This is because the fluid is forced to move across the surface as in the case of a fan.

Radiation refers to the emission of electromagnetic energy. In particular, it refers to infrared photons that carry heat energy. All matters are subject to absorption and emission of some electromagnetic radiation. The net amount of radiation determines whether there will be a heat gain or heat loss.

1.5. OTHER BASIC CONCEPTS OF THERMODYNAMICS

There are some basic concepts which help in understanding the complex ideas presented in this book. They are discussed below.

1.5.1. Temperature

As per energy education, the amount of heat which a substance transfers are determined by the speed and number of atoms and molecules in motion. The temperature is higher when the atoms or molecules move faster, and the quantity of heat transferred is determined by the number of atoms or molecules in motion.

> Temperature is "a measure of the average kinetic energy of the particles in a sample of matter, expressed interms of units or degrees designated on a standard scale," according to the American Heritage Dictionary.

Celsius is the most common temperature scale used. It is based on the freezing and boiling point of water. The freezing point is considered 0°C and the boiling point is considered 100°C.

The Fahrenheit scale (°F) is the other scale used which also derives its assigned values from freezing and boiling pints of water which is 32°F and 212°F, respectively.

Kelvin scale (K) is used widely all over the world by scientists because it works in calculations. It derives its name from William Thomson, 1st Baron Kelvin. The increment in Kelvin is same as Celsius scale. A temperature of 1K is equal to a change of 1°C.

The Kelvin scale starting point is where there is absolutely no heat energy. The absolute zero in Kelvin is the point when all molecular motions stop. The equivalent of 0 K in Celsius and Fahrenheit is minus 273.15°C and 459.67°F, respectively.

1.5.2. Thermal Conductivity

> Thermal conductivity (k) is defined as "the rate at which heat passes through a specified material, expressed as the amount of heat that flows per unit time through a unit are a with a temperature gradient of one degree per unit distance," according to the Oxford Dictionary.

The unit for measuring thermal conductivity is watts per meter per kelvin (W/m·K). Values of thermal conductivity *k* for metals such as silver and copper are 428 W/m·K and 401 W/m·K respectively. Therefore, these materials are considered appropriate for cooling fins of computers and automobile radiators. They can carry heat at a quick rate and exchange it with the environment. Diamond has the highest thermal conductivity amongst all naturally occurring substances at 2,200 W/m·K.

There are some materials which are very poor conductors of heat and considered useful because of this attribute. This property is referred to as thermal resistance or *R*-value. It refers to the rate at which a material transmits heat. Examples of such materials are goose down, rock wool and Styrofoam. They have various applications such as insulation of exterior building walls, winter coats, and thermal coffee mugs. *R*-value is measured in terms of units of square feet times degrees Fahrenheit times hours per British thermal unit (ft^2·°F·h/Btu) for a 1-inch-thick slab.

1.5.3. Newton's Law of Cooling

Sir Issac Newton postulated his Law of Cooling in 1701. The law, in essence, states that the rate at which change of temperature occurs is proportional to the difference between the temperature of the object and the temperature of the surrounding temperature. As a result, an exponential decay occurs in temperature difference. An example of this law is if a warm object is placed in a cold bath, the difference in their temperature will be reduced by half in a certain period of time. In the same time period, the remaining difference in temperature will again reduce by half. This process goes on, until the difference in temperature becomes so small that it is difficult to measure.

1.5.4. Entropy

Heat waste is a part of all thermodynamic systems. As a consequence, the entropy increases for these systems. Entropy for a closed system refers to the American Heritage Dictionary is "a quantitative measure of the amount of thermal energy not available to do work." In a closed system, the entropy *always* increases. It does not decrease at all. Moreover, in moving parts, friction is present which generates waste heat. As a result, radiative heat inevitably comes out from the system.Therefore, some experts believe that the perpetual motion machine (PMM) is impossible. According to Siabal Mitra, a professor of physics at Missouri State University, a machine which is 100 % efficient is cannot build. Hence a machine in perpetual motion cannot be a reality. However, some experts believe that it is possible and continue to try to build these PMMs.

Another way of defining entropy is that it is "a measure of the disorder or randomness in a closed system." This randomness keeps increasing. This can be explained with the help of an example.When hot and cold water is mixed, the large cup of warm water is more disordered than the two small cups of cold and hot water. Energy is required to be added if the two are to be separated again, otherwise separating them is not possible.

A simple example from day to day life, it is not possible to remove cream from coffee or unscramble an egg which has been scrambled.There are some processes which appear as completely reversible ones. However, in practice there are no processes which are completely reversible. Therefore, entropy provides an arrow of time. Increasing entropy has forward direction and vice versa.

1.6. THERMODYNAMIC SYSTEMS

A thermodynamic system refers to a macroscopic region of universe which is being studied. Every other component which lies outside the system is referred to as its surroundings.

There are system boundaries which separate the system from its surroundings. Some boundaries may be movable, while others may be fixed. These boundaries control the transfer between the system and its surroundings. These transfers are mainly mass and energy transfers.

For example, a gas in a cylinder is a thermodynamic system. If the cylinder receives heat from below, the gas is heated, and the piston of the cylinder will rise.

The rise in the piston is accompanied by the movement of its boundary. During this thermodynamic process, heat, and work crosses the boundary of the system, however, the matter within the system can still be identified.

There are broadly three types of thermodynamic systems open system, closed system, and isolated system. Closed system is one in which exchanges of energy occur but not of matter. In open system exchanges of energy and matter occur. Isolated system does not exchange energy or matter.

There are certain concepts associated with thermodynamic systems:

1.6.1. Extensive and Intensive Properties

A quantitative property of a system describes macroscopic features at a given point of time which is fixed. It is important to note that these features may vary with time.

Table 1.2. Symbols and SI Units for Some Common Properties

Symbol	Physical Quantity	SI Unit
E	energy	J
m	mass	kg
n	amount of substance	mol
p	pressure	Pa
T	thermodynamic temperature	K
V	volume	M^3
U	internal energy	J

Source: Table by Chem.libretext.

The symbols of some thermodynamic properties that are discussed in this chapter are mentioned in the table along with the SI units. There are two categories under which most thermodynamic properties will fall, namely extensive and intensive. The two can be distinguished using certain considerations.

Hypothetically, when a system is divided into parts by an imaginary surface, the extensive property of the system is the sum of property of the two parts. In other words, the additive property is extensive. Some examples are mass, energy, volume, amount, and surface area of a solid.

A more restrictive definition of an extensive property is that when intrinsic properties remain constant, the property must not only be additive but also proportional to the mass or amount. As per definition, volume, mass, energy, and amount are extensive, while surface area is not.

When a system is divided into two or more parts of arbitrary size, any property which holds the same value in each part and the whole is referred to as intensive property. Some examples are density, pressure, concentration, and temperature. In a homogeneous region, the intensive property is the same everywhere, but varies from point to point in a heterogeneous region. This is referred to as a local property.

1.6.2. State of System

State of a system is the condition of system which can be described by measurable properties. The properties of a system remain unchanged at a given state. Thus, when the value of even one property changes, the state changes to a different one. When the system is in equilibrium, no unbalanced potential is present within the system. A system which is in equilibrium does not undergo any change when it is isolated from its surroundings. The different types of equilibrium are discussed below:

- Thermal equilibrium: The temperature remains constant throughout the system.
- Mechanical equilibrium: The pressure remains the same throughout the system, however, the pressure may vary within the system due to gravitational effects.
- Phase equilibrium: This term is used in reference to a two-phase system, when the mass of each phase reaches an equilibrium level.
- Chemical equilibrium: The chemical composition of a system does not change with time. In other words, no chemical reactions occur.

1.6.3. Variables, Equation, and Function

State Variables: It refers to macroscopic parameters that explain the properties of the system at an equilibrium state. It includes temperature (T), pressure (P), volume (V). The value of the state variables remains same in every part of the system.

State Equations: It represents the mathematic relations between the state variables. For example: P= P(v, T) or T= T(P, v)

State Function: It refers to those properties whose are determined by the state of the system the path by which the state was reached does not affect the properties. It includes pressure, volume, temperature, internal energy.

1.6.4. Processes in a System

The occurrence of a thermodynamic process changes the system. Types of processes that can occur in a thermodynamic system are the following:

- Isothermal processes are those which occurs at constant temperature;
- Isochoric processes are those occurs at constant volume;
- Isobaric processes are those occurs at constant pressure;
- Adiabatic processes are those occurs without loss or gain of energy as heat;

- Isentropic processes are those occurs at constant entropy; and
- Isenthalpic processes are those occurs at constant enthalpy.

The process may be reversible or irreversible. It may be spontaneous such as the flow of heat from a hot to cold body or non-spontaneous such as flow of heat from a cold to hot body. The process may be cyclic which can be repeated indefinitely without having an effect on the final state of the system. A system is said to be in equilibrium when values are definite and remain unchanged.

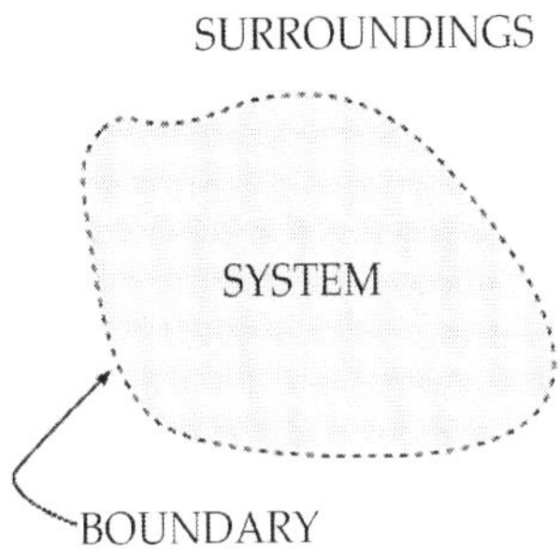

Figure 1.1. *System and its surroundings.*

Source: Image by Wikipedia.

1.6.5. Control Volume

Though the boundary of the system will change with time, the mass remains unchanged. A good example is a piston-cylinder arrangement. The piston is filled with gas and closed by a piston. The gas is the system in the given example. The system can be operated by pushing or pulling the piston to expand or compress the gas. So basically, the boundary of the system is moving. However, the mass remains confined within the boundary as system is a fixed mass as per the definition. The definition does not place a restriction on the work or energy crossing the boundary.

This is an example of a system which is easy to analyze. All systems are not this simple. In fluid dynamics, there are systems which are much more complicated. In fluid dynamics, there are systems which are much more complicated. An example of a complicated system is the flow about the Airplane. A system in such a flow undergoes a lot of changes and hence difficult to analyze.

The boundary keeps changing and goes through unmanageable distortions. The approach to the system is ruled out. Another example of a complex system is turbomachinery, hydraulic systems etc.The term control volume is associated with system. In this approach the focus is on a fixed mass of fluid and instead establish a window in which the flow the observed. This is called control volume. As against the system, a control volume has a fixed boundary. This boundary can be crossed by mass, energy, and momentum. A balance of the mass, energy, and momentum that flow across the boundary is performed. This facilitates in deducing the changes that could occur to properties of flow inside the control volume.

Generally, the shape of the control volume does not change. Hence it is a more convenient approach to visualize volume. Hence, it is the most frequently used approach in fluid dynamics.

Most of the time, a fixed volume control is considered. However, there may be control volumes which change their boundary, or which may deform, etc. As a result, more complicated questions arise. Control surface is the term used to refer to the boundary of control volume.

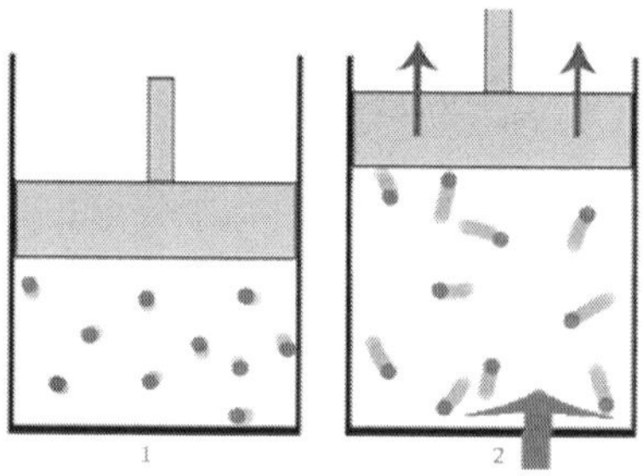

Figure 1.2. *Representation of a piston.*

Source: Image by Wikimedia.

1.7. PROPERTIES OF A PURE SUBSTANCE

It is important to understand the meaning of pure substance to learn about its properties. It refers to a substance which has a homogeneous and invariable chemical composition. The chemical composition of the pure substance remains the same in all phases. For example, water is a pure substance. Ice and vapor have the same chemical composition as water. So does a mixture of water and vapor and water and ice.

It is important to understand that water can exist in many forms. When water is heated it becomes vapor and when cooled it becomes ice. This leads to the concept of different phases of a substance.

A phase can be defined as a stage where the quantity of matter is homogeneous throughout. More than one phase can exist for any matter as seen in the example of water. These phases are separated by the phase's boundary. The pressure and temperature of each of these phases varies. In terms of thermodynamics, it is referred to as states. Each state can be identified or described.

1.7.1. Vapor-Liquid-Solid-Phase Equilibrium

To understand this, an example is considered. A piston cylinder has 1 kg of water and the piston and weight maintain a pressure of 0.1 MPa (Mega Pascals) in the cylinder. The initial temperature is 20°C. The transfer of heat has the following effects:

- It causes an appreciable increase in the temperature;
- There is a slight increase in the specific volume; and
- The pressure is maintained constant.

When the temperature increases to 99.6°C, there is an additional heat transfer which changes the phase. Some of the liquid becomes vapor during which the pressure and the temperature remain constant. However, there is an increase in the volume. After the last drop of liquid has changed to vapor, the transfer of further heat increases both the temperature and specific volume.

Saturation temperature refers to the temperature at which vaporization occurs at a given pressure which is referred to as the Saturation Pressure. The saturation pressure for water at 99.6°C is 0.1 MPa and vice versa. In a pure substance, there is a definite relation between saturation temperature and saturation pressure. A typical curve which is referred to as Vapor Pressure.

A substance which exists in the liquid state at the saturation pressure and temperature is referred to as saturated liquid.

When the temperature of a liquid is lower than the saturation temperature for a given pressure, it is referred to as a Subcooled Liquid.

This implies that at the give pressure, the temperature is less than the saturation temperature. It is also referred to as a Compressed Liquid which implies at the given temperature, that the pressure is greater than the saturation pressure.

Some substances exist as part liquid and part vapor at the saturation temperature. The quality of these substances is defined as a ratio of the mass of vapor to total mass. For example, of the mass of vapor is 0.2 kg and the mass of the liquid is 0.8 kg, quality would be 0.2/(0.2 + 0.8).

The term quality has context only when a substance is at saturated temperature and pressure. Saturated Vapor is the term used to refer to a substance exists at the vapor state in its saturation temperature. The vapor may exist at a temperature greater than the saturation temperature which is referred to as superheated vapor. In a Superheated Vapor, the temperature and vapor are independent properties. Even when the pressure is constant, the temperature is increased.

The critical point is the point at which the saturated vapor and saturated liquid are almost alike. The pressure, temperature, and specific volume at the critical point are referred to as critical pressure, critical temperature and critical volume respectively.

1.7.2. Independent Properties of a Pure Substance

Two independent properties are identified in the case of a simple compressible pure substance. For example, when the temperature and specific volume of superheated steam is given, it is possible to determine the state of steam.

Experimental observation has enabled to be determine the equation for P-v-T Behavior of gases at low density. Equation of state (EOS) for the vapor phase of a simple compressible substance is given below.

$$Pv = \bar{R}T$$

Here,

$$\bar{R} = 8.3144 \frac{\text{kN m}}{\text{kmol K}}$$

It represents the universal gas constant.

The EOS on unit mass basis can be derived by dividing the molecular weight by M

$$Pv = RT$$

This equation of the state can be expressed in terms of total volume for a mass *m* which has a volume *V*.

$$PV = n\bar{R}T$$

$$PV = mRT$$

The equation of the state which is expressed above is known as the ideal-gas EOS. When the density of the gases is low, it approaches ideal gas Behavior. The P-v-T relationship is expressed by the ideal-gas equation of state.

1.8. STEAM CYCLE

A change in state is said to occur when properties of a system change. For example, in the cylinder with piston, when the piston rises the specific volume increases and pressure decreases as a result of which a change in state occurs. The system passes through a succession of states which is referred to as process.

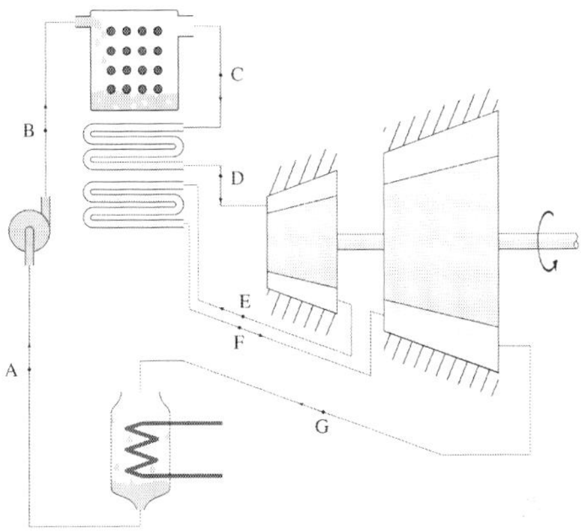

Figure 1.3. *Superheated Rankine cycle.*

Source: Image by Wikipedia.

A system is said to have undergone a cycle when from its initial state, it goes through a series of changes in state or processes and returns to its initial state. The value of the properties at the end of the cycle is the same as in the beginning. For example, steam water circulating through a steam power plant is a cycle.

An elementary steam cycle or the Rankine cycle refers to the fundamental operating cycle of all power plants. It consists of an operating fluid which is evaporated and condensed on a continuous basis. The temperature range determines the operating fluid which will be most suited.

The pressure-enthalpy (p-h) and temperature-entropy (T-s) diagrams of this cycle are given in Graph below.

1.8.1. Steps in a Steam Cycle

The following are the steps involved in a Rankine cycle:

1–2-3 Isobaric Heat Transfer: High pressure liquid enters the boiler and is heated. A feed pump is present through which the liquid enters. The liquid reaches the saturation temperature. It is heated further which leads to the evaporation of the liquid. It fully converts to saturated steam.

3–4 Isentropic Expansion: Inside the turbine, the vapor gets expanded. It produces work in the process which is converted to electricity. The temperature of the cooling medium limits the expansion in the practice. As the process moves into the two-phase region, the erosion of the turbine blades by liquid entrainment in the vapor stream also limits the temperature. An exit vapor is produced whose qualities should be greater than 90%.

4–5 Isobaric Heat Rejection: A vapor-liquid mixture is emitted from the turbine. This is condensed at a low pressure. This takes place in a surface condenser and is facilitated by cooling water. Ideally the pressure of vapor is way below the atmospheric pressure in case of well-designed and maintained condensers.

5–1 Isentropic Compression: In a feed pump, the pressure of the condensate is increased. The pump work is relatively small, and this can be attributed to the low specific volume of liquids. Hence it is often not considered for thermodynamic calculations.

Rankine Cycle

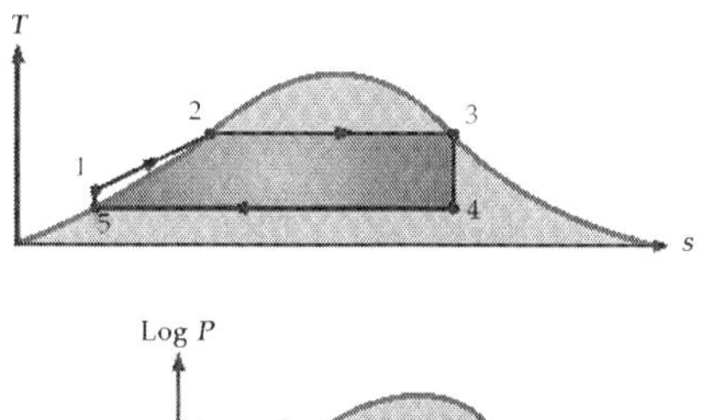

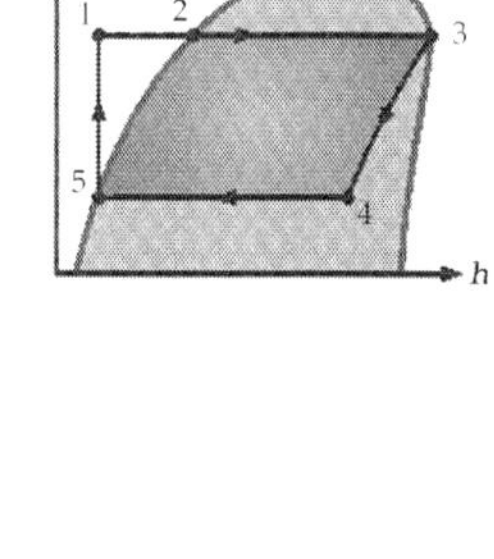

T-s and p-h diagrams

1.8.2. Efficiency and Inefficiency of Rankine Cycle

The following formula calculates the efficiency of power cycles:

$$\eta = \frac{W_{net}}{Q_{in}} = \frac{W_{34} - W_{51}}{Q_{13}}$$

The First Law of Thermodynamics helps in calculating the values of heat and work. The assumption of isentropic expansion determines the steam quality x at the turbine outlet i.e.

$$s_3 = s_\pi^*(p_3) = x(s_l^*(p_4)s_\pi^*(p_4))$$

In this formula s_π^* represents the entropy of vapor and Si* represents the entropy of liquid.

The efficiency of the ideal Rankin cycle described above is close to Carnot efficiency. In real plants, an irreversible process is associated with every stage of Rankine cycle. This reduces the overall efficiency of the Rankine cycle.

In the calculation of the overall cycle efficiency, turbine, and pump irreversibilities can be included. This is can be achieved by defining a turbine efficiency according to the following formula.

$$\eta_t = \frac{W_a}{W_s} = \frac{(h_{t,in} - h_{t,out})_a}{(h_{t,in} - h_{t,out})_s}$$

In this formula, the subscript act indicates actual values and subscript indicates isentropic values and a pump efficiency

$$\eta_p = \frac{W_s}{W_a} = \frac{V(p_1 - p_6)}{W_a}$$

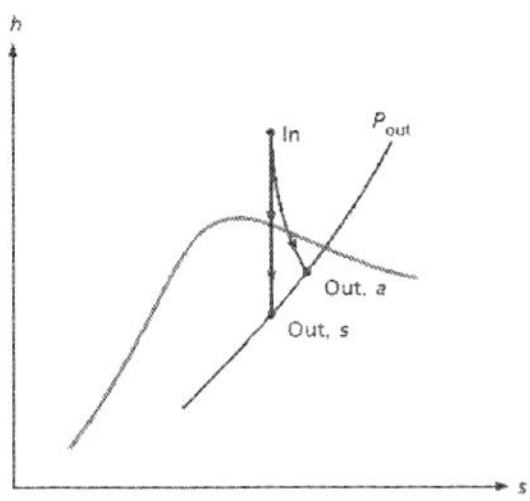

Turbine efficiency

When η_t and η_p are known, the values for the isentropic processes can be used to determine the actual enthalpy after the compression and expansion steps. As a result of turbine efficiency, the work produced in the turbine reduces. This has an impact on the overall efficiency.

The enthalpy of the liquid which leaves the pump increases due to the inefficiency of the pump. As a consequence, the amount of energy which is needed to evaporate the liquid is reduced. Nevertheless, the energy expended by the boiler is less compared to the energy requirement of the pump.

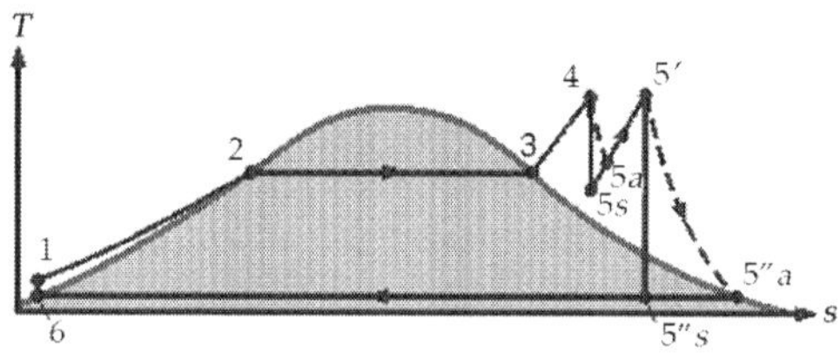

Rankine cycle with vapor superheating

The conversion of fuel energy into useable steam energy is low in boilers. Even the most advanced boilers only transform 40% of fuel energy. This wastage can be attributed to two causes:

- The combustion gas temperature is quite high and ranges between 1000°C and 2000°C. This temperature range exceeds the highest vapor temperature. The entropy increases due to the transfer of heat across a large temperature difference.
- The combustion or oxidation process which occurs at a technically feasible temperature is highly irreversible.

The heat transfer surface in the condenser has a finite value. Therefore, condensations take place at a temperature which is higher than the temperature of the cooling medium. Entropy is again generated due to heat transfer across a temperature difference. Dirt gets deposited in the condensers during the operation the cooling water. This has an impact on the efficiency.

1.8.3. Increasing the Efficiency of Rankine Cycles

The efficiency of the Rankine cycle can be increased in the following ways.

Pressure Difference

The network produced in the Rankine cycle is given by the area of cycle process in the figure above. Increasing the pressure in the boiler along with reducing the pressure will increase the network.

Superheating and Reheating

A process should take place at a temperature which is proximate to the temperatures of the high temperature and low temperature reservoirs. This reduces the irreversibility of any process. In order to achieve this the condenser has to operate at subatomic pressure. The saturation pressure is a factor which limits the temperature in the boiler. If the saturated vapor is superheated, a further increase in temperature takes place.

As a result of this, the vapor quality increases after the turbine thereby lowering the erosion of turbine blades. The vapor is reheated after it expands in the high-pressure turbine. This is again reheated in another low-pressure turbine.

Feed Water Preheating

The feed pump emits cold liquid which is mixed with the saturated liquid in the boiler. Another technique is to re-heat it to the boiling temperature. This leads to an irreversible process which lowers the efficiency of the boiler.

The Carnot process states that when heat transfer occurs isothermally, the highest efficiency is achieved. Bleed vapor from various positions of turbine is passed through external heat exchangers to preheat the feed liquid. It is desirable to keep the temperature of the bleed steam close to the temperature of the feed liquid.

Combined Cycles

Better utilization of high temperature in which the fuel is combusted occurs if a gas turbine or Brayton engine is used along with a Rankine cycle. When this occurs, the hot gas emitting from the turbine is used to fuel the boiler.

A co-generation system helps improve energy efficiency. It is a system that provides both electrical power and heat. When used in a Rankine cycle, the energy rejected is utilized for space heating, process steam and other applications involving low temperatures.

1.9. THE CARNOT CYCLE

Nicolas Leonard Said Carnot, a French mechanical engineer in the French Army, came up with a model for heat engine. The principle on which he based was named Carnot Cycle. This cycle mainly emphasizes on the relationship between the volume, pressure, and temperature of gases. It is also concerned with how input energy changes form and can perform work outside the system.When a gas is compressed, its temperature rises. Therefore, it becomes hotter than its environment. The heat can then be extracted using a heat exchanger from the hot gas. When it is allowed to expand, the gas cools down. This principle is used in heat pumps which has its application in air conditioners, heaters, and refrigerators.

When a gas is heated, its pressure increases and makes it expand. This expanded pressure is employed in driving a piston. Heat energy is converted to kinetic energy in the process. The functioning of heat engines can therefore be explained using the Carnot Cycle. It is discussed in detail in a later chapter.

1.10. CONCLUSION

Thermodynamics is an interesting branch in physics which is concerned with the relation between heat and other form of energy. It is a complex science which is concerned with the measurement of thermal energy which a matter or system possess due to its vibrating molecules. Thermodynamics has evolved considerable since the 1600s when the basic concept of heat and motion was established till the 20th century which was marked by the discoveries of eminent scientists. There are some important concepts in thermodynamics besides heat and work, namely, temperature, entropy, and thermal conductivity.

The thermodynamic system is the macroscopic universe which is studied in thermodynamics which is separated from its surroundings by its system boundaries. Pure substances to a substance which has a homogeneous and invariable chemical composition. The chemical composition of a pure substance remains the same in all phases or states of matter.

Two important cycles in thermodynamics are the Rankine cycle and the Carnot cycle. When a system undergoes a cycle, it starts from its initial stage and passes through a series of changes in state or processes and returns to its initial state.

REFERENCES

1. Basic Concepts of Thermodynamics, (2020). [ebook] Available at: http://www.idc-online.com/technical_references/pdfs/mechanical_engineering/Basics_of_Thermodyanamics.pdf (accessed on 11 March 2020).
2. Chemistry LibreTexts, (2019). *2.1 The System, Surroundings, and Boundary*. [online] Available at: https://chem.libretexts.org/Bookshelves/Physical_and_Theoretical_Chemistry_Textbook_Maps/DeVoe's_%22Thermodynamics_and_Chemistry%22/02%3A_Systems_and_Their_Properties/2.1_The_System%2C_Surroundings%2C_and_Boundary (accessed on 11 March 2020).
3. Energyeducation.ca. (n.d.). *Heat vs work—Energy Education*. [online] Available at: https://energyeducation.ca/encyclopedia/Heat_vs_work (accessed on 11 March 2020).
4. Eoht.info. (n.d.). *Branches of Thermodynamics—Hmolpedia*. [online] Available at: http://www.eoht.info/page/Branches+of+thermodynamics (accessed on 11 March 2020).
5. Introduction to Thermodynamics, (2020). [ebook] Available at: https://canteach.candu.org/Content%20Library/19930202.pdf (accessed on 11 March 2020).
6. Lucas, J., (2015). *What Is Thermodynamics?*[online] livescience.com. Available at: https://www.livescience.com/50776-thermodynamics.html (accessed on 11 March 2020).
7. Powers, J. M., (2020). *Lecture Notes on Thermodynamics*. [ebook] Department of Aerospace and Mechanical Engineering University of Notre Dame. Available at: https://www3.nd.edu/~powers/ame.20231/notes.pdf (accessed on 11 March 2020).
8. Mefanet, C., (n.d.). *Thermodynamic System/Description – WikiLectures*. [online] Wikilectures.eu. Available at: https://www.wikilectures.eu/w/Thermodynamic_system/description (accessed on 11 March 2020).
9. Muller-Steinhagen, H., (n.d.). Rankine cycle. *A-to-Z Guide to Thermodynamics, Heat and Mass Transfer, and Fluids Engineering*.
10. Web.mit.edu. (n.d.). *1.2 Definitions and Fundamental Ideas of Thermodynamics*. [online] Available at: https://web.mit.edu/16.unified/www/FALL/thermodynamics/notes/node11.html#SECTION04124000000000000000 (accessed on 11 March 2020).

11. Wolframscience.com. (2020). *Historical Notes from Stephen Wolfram's A New Kind of Science*. [online] Available at: https://www.wolframscience.com/reference/notes/1019b (accessed on 11 March 2020).
12. Www-mdp.eng.cam.ac.uk. (n.d.). *System and Control Volume*. [online] Available at: http://www-mdp.eng.cam.ac.uk/web/library/enginfo/aerothermal_dvd_only/aero/fprops/cvanalysis/node10.html (accessed on 11 March 2020).

CHAPTER 2

Laws of Thermodynamics and Thermodynamic Potentials

CONTENTS

Thermodynamics basically deals with the relation between heat and other form of energy. The Laws of thermodynamics are important for understanding the thermodynamic principles. This chapter involves the study of the laws of thermodynamics that includes Zeroth law, first law, the second law and the third law. The chapter provides insights about the four potentials of the thermodynamics. The thermodynamic potentials are the measuring scale that measures the thermodynamic state of a system.

2.1. INTRODUCTION

According to the Agency, Thermal energy is the energy a substance or system has due to its temperature, i.e., the energy of moving or vibrating molecules.

The study of energy interactions between systems and the effect of these interactions on the system properties is known as Thermodynamics. There is a transfer of energy between systems that takes place in the form of heat or work. Thermodynamics is a subject that deals with system equilibrium.

A thermodynamic system is defined as the study of quantity of fixed mass and identity upon which attention is focused. Thermodynamic system is the everything this chapter needs to study.In thermodynamics, a system could be as simple as gas in a cylinder and as complex as a nuclear power plant. Things that are external to the system is the surroundings. There is a boundary that separates system.The thermodynamic system is classified into three different systems, i.e., closed systems, open systems, and isolated systems. There is a specified region in space where most of the attention is focused, is known as control volume. The control volume can be considered as an open system. The control surface separates the surroundings from control volume. This results in easy entry and exit of the mass and energy in the control room.

There involves a wide variety of processes in refrigerating and air conditioning. The understanding of the nature of the process path is important since heat and work depends on the path. In this, the process can be said to have different states that occurs within the initial state to the final state. In thermodynamics, the process is said to have completed a cycle if the process starts from the initial state and finally arrives at the initial state. According to the website of the Energy Education of the Texas Education Agency, Thermal energy is the energy a substance or system has due to its temperature. A professor of physics at Missouri Southern State University, David McKee has defined thermodynamics that involves measuring this energy, which can be "exceedingly complicated." Because there are large numbers of atoms and molecules that interact in a complicated way. These atoms and molecules can be described with a very small number of measurements or numbers if equilibrium is achieved. This is known to be the mass of the system, the pressure of the system, and the volume of the system, or some other equivalent set of numbers.

2.2. FUNDAMENTAL LAWS OF THERMODYNAMICS

The Classical thermodynamics is based on four fundamental laws called zeroth, first, second, and third laws of thermodynamics. These laws are important for understanding the thermodynamic principles as these laws defines the thermodynamic properties.

Temperature is defined by Zeroth law, first law defines internal energy, the second law defines entropy and the third law is used to obtain the absolute entropy values.

These laws are not mathematical driven formulas rather they depend on the natural phenomena of the human observation. Also, there are no exception to these observations, these are accepted as laws. Conservation of mass is a fundamental concept. It states that mass can neither be created nor destroyed.

2.3. THE FOUR LAWS OF THERMODYNAMICS

Initially, three laws were expressed by the fundamental principles of thermodynamics. Later, one more law was found apparently because it had seemed so obvious that it did not need to be stated explicitly. In order to complete the set of rules, a new law was also included in the fundamental laws of thermodynamics.

However, the scientists came across a problem that was regarding the numbering of the new law. It was because the first three laws were already numbered in series of their order. The problem came with the new law with regards to assign the number. The scientists could not renumber the existing laws as it would have created confusion.

Also, placing the pre-eminent law at the end of the list of the laws would have made no sense. Therefore, a British physicist, Ralph H. Fowler, came up with an alternative idea that solved the dilemma. As per the physicist, the new law was named the "Zeroth Law." These laws are discussed in brief:

2.4. THE ZEROTH LAW OF THERMODYNAMICS

It states that if two bodies are in thermal equilibrium with some third body, then they are also in equilibrium with each other. The law establishes temperature as a fundamental and measurable property of matter.

There are four laws of thermodynamics. One of the laws is the zeroth law of thermodynamics. According to the law, if two systems are in thermal equilibrium with a third system, then the first two systems are in thermal equilibrium with one another.

The fundamental quantities in the Zeroth Law of Thermodynamics are temperature and heat. Under this law, temperature sets the level of heat. Heat is the energy that passes from one body to another through a thermal flow process due to a temperature difference.

Temperature is one of the following basic physical properties, length, mass, time, electric current, and luminous intensity-in terms of which all the other physical properties are defined and the thermal equilibrium of two systems are defined.

As per the law, the lower limit of temperature is –273.15°C, which is not accessible according to the law. However, theoretically, there is no upper limit to the temperature. According to Aristo, the origin of the word is from the Latin word “temperamentum,” which indicates the four basic properties of matter: heat, cold, humidity, and dryness.

It was for distinguishing between the level of heat and its quantity, probably, El-Kindy, in the 9th century. The concept of temperature was ambiguous until the 17th century. Since no methods were known for the quantitative determination of the level of heat till the time.

At that time, the first instrument man’s first instrument for sensing hot and cold was his hand. However, it was soon realized that the sensing by hand is inadequate. After that, the Italian scientist Galileo Galilei (1564–1642), was the first to construct a thermometer. While at that time, other scientists including Santorio (Italian), Robert Fludd (English), and Cornelis Drebbel (Dutch) worked independently on the similar devices.

In 1694, Pavia C. Renaldini, was the first who used melting ice and boiling water as constant points of the temperature scale. Corresponding to the practical scale, Andres Celsius divided the interval between the above points in a mercury thermometer to 100 degrees in 1742.

There are two scales: the Celsius, or centigrade scale (°C) and the Fahrenheit scale (°F). The other scale also prevails that is the thermodynamic scale, which assumes the existence of an absolute scale of temperature, the Kelvin scale (K) or the Rankine scale (°R), which have a lower limit, the absolute zero, 0 K.

2.4.1. Applications of the Zeroth Law of Thermodynamics

In everyday life, the uses of the zeroth law are seen in many activities, some of the applications of the law are:

- The thermometer is the well-known example of the zeroth law of the thermodynamics. Let's say for example the thermostat in a bedroom reads 67 degrees Fahrenheit. That means the room is in thermo equilibrium with the thermostat. The application of the zeroth law to this would state that every other thing present in the room may be wall clock, decorative piece is also at the67-degree Fahrenheit.
- The other example of the zeroth law of thermodynamics is if a glass of ice water and a glass of hot water is placed on the countertop of the kitchen for a few hours, eventually both the glasses will reach thermal equilibrium with the room. The zeroth law when applied state that the room will be in equilibrium with the glasses.
- When a packet of meat is placed in the freezer and left overnight. One can assume that the meat has reached the same temperature as the freezer and the other items present in the freezer.

2.5. THE FIRST LAW OF THERMODYNAMICS

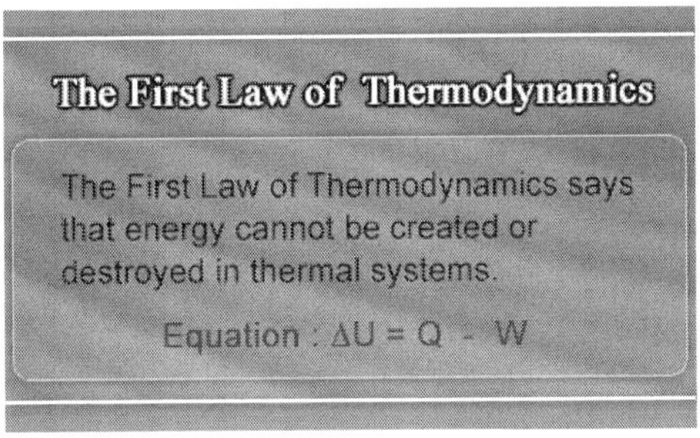

Figure 2.1. *First law of thermodynamics.*

Source: Image by Flickr.

According to the first law of thermodynamics, the total increase in the energy of a system is equal to the increase in thermal energy plus the work done on the system. This states that heat is a form of energy and is subject to the principle of conservation.

As per the first law of thermodynamics, the energy can be converted from one form to another with the interaction of heat, work, and internal energy. It also states that the energy cannot be destroyed or created. The first law is the statement of law for conserving the energy.

The law says that heat and work are interchangeable. If there is any disturbance in the system that destroys or creates the energy is known as a Perpetual Motion Machine (PMM) of first kind.

Mathematically, the first law of thermodynamics is represented by:

$$\Delta U = q + w$$

where, ΔU is the total change in the internal energy of a system; q is the heat that is exchanged between a system and its surroundings; and w is the work done by or on the system.

In the first law of thermodynamics, when a system is undergoing a cyclic process, the law is represented by:

$$\oint \delta Q = \oint \delta W$$

where, $\int \delta Q$ = net heat transferred during the cycle; $\int \delta W$ = network transferred during the cycle.

The above equation can be written as:

$$\int (\delta Q - \delta W) = 0$$

From the above equation, $(\delta Q - \delta W)$ must be a point function or the property of the system. The internal energy is represented by, U. Therefore, Mathematically, internal energy can be written in equation as:

$$dU = \delta Q - \delta W$$

The internal energy of a system is the sum-total of all the forms of energy. The energy can be thermal, molecular, lattice, nuclear, rotational, vibrational, etc.

2.5.1. First Law of Thermodynamics for a Closed System

Let us assume that the internal energy of a closed system at an equilibrium state 1 be U_1. The amount of heat transferred across the boundaries be ${}_1Q_2$ and the amount of work done by the system be ${}_1W_2$. The system then comes to an equilibrium state 2.

The integrated equation is written as,

$$U_2 - U_1 = {}_1Q_2 - {}_1W_2$$

Mass of the system is m and u denotes the specific internal energy of the system then, the equation can be written as,

$$m(u_2 - u_1) = m({}_1q_2 - {}_1w_2)$$

$$\text{or, } u_2 - u_1 = {}_1q_2 - {}_1w_2$$

where, ${}_1q_2$ is the heat transferred per unit mass of the system; and ${}_1w_2$ is the work done per unit mass of the system.

Flow Work

The fluid enters and leaves the system in an open system some matter. The fluid is required to enter the system against the system pressure. At the same time, flow work is needed to expel the fluid from the system. For the above process, flow work is required.

From above, it can be shown that the flow work is given by the product of pressure, p, and a specific volume, v. The flow work can be written as:

Flow work = pv

Enthalpy

While the analysis of open systems, it becomes necessary to combine the specific flow work 'pv' with internal energy 'u' since both the internal energy and the flow work will increase the energy of the system. The total sum of specific internal energy and specific flow work is an intensive property of the system and it is called specific enthalpy, h. Therefore, specific enthalpy, h is given by:

$$h = u + pv$$

2.5.2. First Law of Thermodynamics for Open System

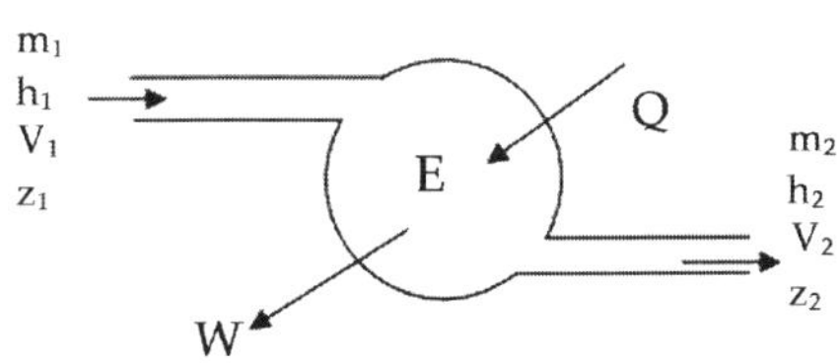

The above figure represents an open system where, m_1 and m_2 are the mass flow rates at inlet and outlet and h_1 and h_2 are the specific enthalpies at inlet and outlet. The inlet and outlet velocities are V_1 and V_2 and z_1 and z_2 are the heights at inlet and outlet with reference to a datum.

The rate of heat transferred is q and the rate of work transferred is w to the system. The total energy of the system is represented by E.

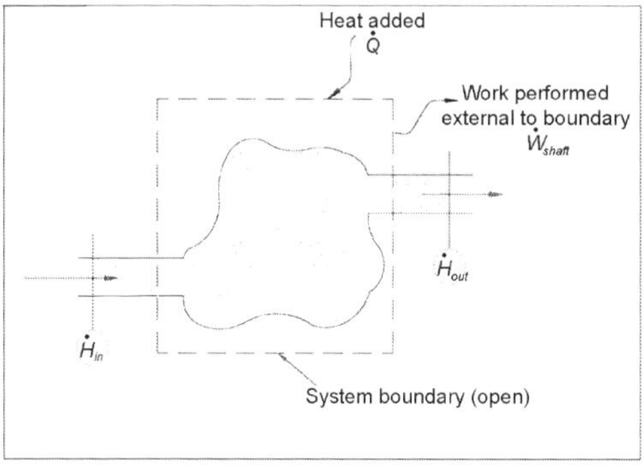

Figure 2.2. *First law in open system.*

Source: Image by Wikimedia Commons.

Then, the first law of thermodynamics for the open system is given by:

$$\frac{dE}{dT} = m_2(h_2 + \frac{V_2^2}{2} + gz_2) - m_1(h_1 + \frac{V_1^2}{2} + gz_1) + W - Q$$

where, (dE/dt) is the rate at which total energy of the system changes; and g is the acceleration due to gravity.

2.5.3. First Law for Open System in Steady State

The first law of thermodynamics in a steady state, says that the time rate of change of all the quantities is zero, and mass is also conserved. Therefore, there will be no change in the total energy and the mass with time. Hence, it can be said that (dE/dt) is zero and from conservation of mass, $m_1 = m_2 = m$.

The first law of thermodynamics in a steady state is represented by:

$$(h_2 + \frac{V_2^2}{2} + gz_2) - (h_1 + \frac{V_1^2}{2} + gz_1) = q - w$$

In the equation, q is the specific heat transfer rate and w is the specific work transfer rate.

2.5.4. Examples of First Law of Thermodynamics

2.5.4.1. Example – Energy Flow in a Diesel Engine

There are different types of fuel that have different amounts of energy. In a gallon or liter of fuel, there is a fixed amount of energy. The first law of thermodynamics applies in the diesel engine. When an engine burns fuel, the energy stored in fuel's chemical bond is converted into useful mechanical work and finally into heat. There is no fuel or energy except the energy created. There is no other magical device that can change the amount of energy in the device.

According to the first law, when all the energy of the fuel is released by burning in the engine's cylinder, the energy does not disappear. As per law, the energy remains the same and must be accounted for. The first law defines the conservation of energy principle.

In case of the diesel engine, the energy is converted into thermal energy (heat) or mechanical energy (work). Every 100 units of the fuel energy burned in the engine; 100 units of converted energy has to end up somewhere but not disappear.

Now, the sum of all the energy out will be equal to the sum of energy into the engine. This is what the first law of thermodynamics predicts. There are chances of human errors if the result does not look the same.

2.6. THE SECOND LAW OF THERMODYNAMICS

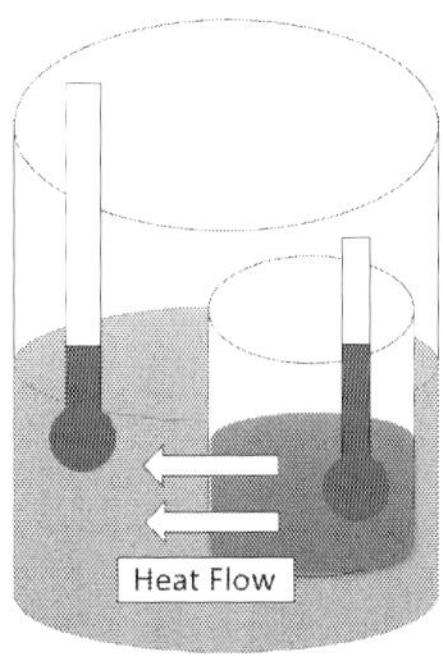

Figure 2.3. *Second law of thermodynamics.*

Source: Image by Wikipedia.

The second law states that the heat energy cannot be transferred from a body at a lower temperature to a body at a higher temperature without the addition of energy. This is the reason why it costs money to run an air conditioner.

According to the second law of thermodynamics, the state of entropy of an entire universe of an isolated system increases over time and can never be negative. This law is the limit law. It gives the upper limit of the efficiency of the system. This law states that the process follows in a certain direction and not in opposite direction.

The law also defines the important property called entropy. The heat does not flow spontaneously from a body at lower temperature to a body at higher temperature. In order to transfer heat from lower temperature body to higher temperature body continuously, i.e., to maintain the low temperature, there is a need of a system that requires work input from external source. The system of refrigeration can solve this problem. This is one of the principles of the second law of thermodynamics, which is known as Clausius statement of the second law.

As per the second law, there are certain things that happens in one direction and not the other. This is known as the arrow of time. The thermodynamic arrow of time, i.e., entropy measures the disorder within the system. It is denoted by ΔS, the change in entropy suggests that system will become disordered as time increases.

2.6.1. Clausius' Statement of Second Law

According to Clausius' Statement of Second Law, it is important to transfer heat from low temperature to high temperature without work from external source. The cyclic process should be followed. All the energy that is transferred to a system as work is dissipated as heat transfers. The energy supplied as heat transfer cannot be continuously converted into work with given thermal efficiency of 100%, rather only some part of it can be converted into work. However, the remaining part is to be rejected at a low temperature in the surroundings.

If it would have been possible to obtain work continuously by transferring heat given a single heat source, then automobiles would run by deriving energy from atmosphere at no cost. There is a hypothetical machine that can achieve it is called Perpetual Motion Machine of second kind. This fact is explained in Kelvin-Planck Statement of the Second law.

2.6.2. Reversible and Irreversible Processes

If the systems and surroundings can be restored to their respective positions by reversing the direction of the process or by reversing the flow of heat transfer and work transfer, the process is said to be reversible with respect to system and surroundings. This system is irreversible if the criteria is not fulfilled.

The work done in the presence of friction, by moving the piston in a cylinder. In such case, a part of the work is dissipated as heat and it cannot be recovered fully if the direction of process is reversed. In similar case to this, the heat transferred through a temperature difference from higher temperature to a lower temperature, the direction of the eat transferred cannot be reversed because in order to transfer heat from lower to higher temperature, there would be requirement of external work output.

Reversible process is called a hypothetical process in which work is done in the absence of friction and heat transfer occurs isothermally. In irreversible process, the loss in work output and loss in availability and useful work.

2.6.3. Heat Engines, Refrigerators, Heat Pumps

A heat engine is an important part in thermodynamic cycle. Heat engines helps in transferring heat from low temperature body to high temperature body by doing a certain amount of work done. Heat engine is a device that operates in a thermodynamic cycle. The example of heat engine is a steam power plant.

A refrigerator can be said to be a device that operates in a thermodynamic cycle. It transfers a certain amount of heat from a body at lower temperature to a body at higher temperature by using some external amount of work done. Examples in day to day life are domestic refrigerators and room air conditioners. In the case of refrigerator, the heat extracted from the low temperature body is the required output. A heat pump is similar to refrigcrator, but in this case the heat rejected due to high temperature body is the required output.

2.6.4. Kelvin-Planck Statement of Second Law

Kelvin stated that "it is impossible to convert heat completely in a cyclic process." This implies that there is no other way to convert all the energy of the system into work without losing energy. According to the scientist, to construct a device that operates in a cycle produces no effect other than extraction of heat from a single reservoir and convert all of it into work. Mathematically, Kelvin-Planck statement of law can be represented as: $W_{cycle} \leq 0$ (for a single reservoir).

2.6.5. Carnot's Theorems for Heat Engines

Theorem 1: As per theorem 1 of Carnot, it is impossible to construct a heat engine operating between two thermal reservoirs. This is more efficient than a reversible engine operating between the same two reservoirs.

Theorem 2: The thermal efficiency of all the reversible heat engines operating between the two thermal reservoirs is same. The second law of thermodynamics helps in proving and explaining these two theorems. Just like heat engines, Carnot's theorems can also be formed for refrigerators.

Carnot Efficiency: According to Carnot's theorems, there is maximum possible efficiency for any given two thermal reservoirs. Carnot efficiencies are the efficiencies of completely reversible cycles operating between two thermal reservoirs

Thermal efficiency for a heat engine, η_{HE} is represented as:

$$\eta_{HE} = \frac{W_{cycle}}{QH} = 1 - \frac{Q_C}{Q_H}$$

where, W_{cycle} is the net-work output, Q_C is the heat rejected to the low temperature reservoir and Q_H is the heat added (heat input) from the high temperature reservoir.

Therefore, according to Carnot's theorems, for a reversible cycle $\frac{Q_C}{Q_H}$ is a function of temperatures of the two reservoirs only, i.e.,

$$\frac{Q_C}{Q_H} = \phi(T_C, T_H)$$

As per the absolute Kelvin temperature scale:

$$\frac{Q_C}{Q_H} = \frac{T_C}{T_H}$$

$$\text{hence, } \eta_{Carnot,HE} = 1 - \frac{Q_C}{Q_H} = 1 - \frac{T_C}{T_H}$$

The efficiency of a heat pump and refrigerator is called Coefficient of Performance (COP). Similarly, to heat engines, Carnot's COP for heat pump

(COPHP) and refrigerators (COPR) can be written as:

$$COP_{Carnot,HP} = \frac{Q_H}{W_{cycle}} = \frac{Q_H}{Q_H - Q_C} = \frac{T_H}{T_H - T_C}$$

$$COP_{Carnot,R} = \frac{Q_C}{W_{cycle}} = \frac{Q_C}{Q_H - Q_C} = \frac{T_C}{T_H - T_C}$$

where, W_{cycle} is the work input to the reversible heat pump and refrigerator; Q_H is the heat transferred between the system and the hot reservoir; Q_C is the heat transferred between the system and cold reservoir; T_H is the temperature of the hot reservoir; and T_C is the temperature of the cold reservoir.

2.6.6. Clausius Inequality

The mathematical form of second law of thermodynamics for a closed system undergoing a cyclic process is called Clausius inequality. The equation is represented as:

$$\oint \left(\frac{\delta Q}{T}\right)_b \leq 0$$

As per the above equation, δQ is the heat transfer at a specific part of the system boundary during a portion of the cycle, T represents absolute temperature at that part of the boundary. The subscript "b" provides a reminder that the integrand is calculated at the boundary of the system that executes the cycle.

The equality is applied when there is no internal irreversibility as the system executes the cycle, while inequality applies when internal irreversibility's are present. This inequality provides a ground for the understanding of entropy that can be related to second law of thermodynamics. This is explained in detailed in the further chapters.

2.6.7. Examples of the Second Law of Thermodynamics

- The second law of thermodynamics can be applied in running or powering a car, where higher concentration of energy is useful than lower concentration of energy. The higher concentration of energy is more efficient than lower concentration of energy.
- There are food items like carbohydrates and fats, and liquid fuels like gasoline, petrol have highly concentrated potential energy stored in their chemical bonds. They carry a lot of energy in a small space and it becomes essential for keeping the bodies and

our machines going.

- When a rock is dropped on the ground and it hits the ground. The initial potential energy of the rock gets converted into kinetic energy as and when the rock falls. When the rock hits the ground the energy of the rock is transformed into internal energy of the rock and the molecules present in the ground move faster and the temperature rises slightly. This is due to the application of the second law of thermodynamics.

2.7. THE THIRD LAW OF THERMODYNAMICS

As per the third law, the entropy of a pure crystal at absolute zero is zero. Entropy is sometimes called waste energy, i.e., energy that is not able to do work. There can be no waste energy if there is no heat energy at absolute zero. Entropy is a measure of the disorder in a system. A positive value of the temperature means there is motion within the crystal, which causes disorder to the perfectly placed crystal. Due to these reasons, entropy always has a positive value. There can be no physical system with lower entropy.

The third law of thermodynamics states that the absolute zero of the entropy cannot be achieved. This law defines the absolute value of the entropy. The law also states that "the entropy of perfect crystals is zero at absolute zero." The statement was given by Plank. The concept of entropy is the measure of the disorder of the system.

As per the law, if the probability of achieving a particular state out of a large number of states is 'ω,' then the entropy of the system is equal to $\ln(\omega)$. There is a uniquely defined positions of the atoms and also the transitional movement of molecules ceases at absolute zero.

Additionally, a perfect crystal has all of its atoms alike and the positions of the crystals can be interchanged without changing the state. The probability of achieving this state is unity, that is $\omega = 1$ and $\ln(\omega) = \ln(1) = 0$.

In case, there is an imperfect crystal, there is some entropy associated with configuration of molecules and atoms. The entropy in this case does not tend to zero as $T \rightarrow 0$, but it tends to a constant called the entropy of configuration, even when all motions cease.

The third law of thermodynamics allows the determination of the absolute zero with zero entropy as a reference state. In case of refrigerators, entropy changes are useful rather than the absolute entropy. Hence, entropy is taken a zero or constant at any suitably chosen reference state.

The law also states that the absolute zero cannot be achieved. In order to achieve the absolute zero, the magnetization in aligning the molecules is done. This is followed by cooling and then demagnetization, which helps in the extraction of the energy from the substance and as a result reduces its temperature.

In the process there are infinite number of cycles to achieve the absolute zero.

2.7.1. Importance of the Third Law of Thermodynamics

The third law of thermodynamics is rarely applicable to the day-to-day lives of the people and it governs the dynamics of objects at the lowest known temperatures. Perfect crystal is defined by the law whose atoms are glued in the positions. Therefore, the perfect crystal possesses absolutely no entropy, which is only achievable at the absolute temperature.

The concept of entropy is popular in some other theories as well. While defining the continuous flow of time objectively, as in the case of linear increase in the entropy of the universe.

At 0 Kelvin, ideally the entropy changes for reactions regarding the formation of matter will be zero. Owing to the presence of the tiniest amount of heat, all matter manifests some amount of entropy practically. The coldest measure of the entropy is 3 K, in the distant depths of the Universe, beyond stars and galaxies.

2.8. THERMODYNAMIC POTENTIALS

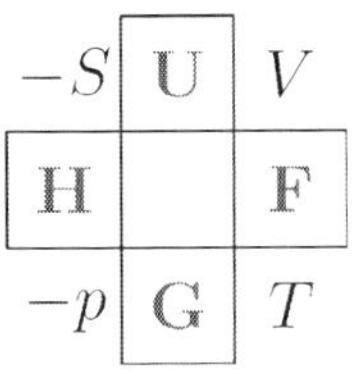

Figure 2.4. *Thermodynamic potentials.*

Source: Image by Wikimedia commons.

Thermodynamic potentials are the scalar quantity used to measure the state of thermodynamics of the system. In a non-isolated system, which is in thermal contact with its surroundings (EXT), at temperature T_{EXT}, having a large heat capacity such that with no change in the T_{EXT}, heat flows from the surroundings into the system.

According to the second principle, it states that the total change of entropy for the system combined with its surroundings is given by δS_{TOT} $\delta S + \delta S_{EXT} \geq 0$ with $\delta S_{EXT} = -\delta Q/T_{EXT}$.

A considered system having a constant volume, $\delta U = \delta Q$. Thus, the equation formed is $\delta (U - T_{EXT}S) \leq 0$. The spontaneous change is due to decrease of $U\text{-}T_{EXT}S$.

Thus, there is a new thermodynamic equilibrium ($T = T_{EXT}$) and it can be achieved for the minimum of the (Helmholtz) free energy of the system $F = U - TS$. Similarly, if a system is in thermal contact with its surroundings and it is held at a constant pressure, the Gibbs free energy $G = U - TS + pV$ will tend to be at minimum at thermodynamic equilibrium.

There are thermodynamic potentials of free energy and the Gibbs free energy. Thermodynamic potentials show some specific functions of state variables, the natural variables of the potential. By keeping these natural variables at a constant rate, and the potential is diminishing, then a closed thermodynamic system evolves towards equilibrium.

When the thermodynamic equilibrium is achieved, it always corresponds to a minimum of one of the thermodynamic potentials. The internal energy $U = f(S,V)$ is such a thermodynamic potential of natural variables S and V. In a mechanical system where there is no heat exchange, an increase of energy is the result of a gain of mechanical work which is the product of a force by a small displacement.

This concept can be generalized in a way that the addition of several products of generalized thermodynamic forces which induce thermodynamic displacements from equilibrium is due to the increase in energy of a thermodynamic system.

An intensive variable drives the thermodynamic force which is conjugated to an extensive variable representing the thermodynamic displacement, p, and V are the conjugated variables, as are T and S. When S and V are kept constant, U will tend to be minimum as the considered closed system tends to thermodynamic equilibrium. The other thermodynamic potentials tending to a minimum for a closed system at equilibrium are:

The enthalpy, if p and the extensive parameters such as S are held constant,

$$H = U + pV \Rightarrow dH = T\,dS + V\,dp$$

the Helmholtz free energy, if T and the extensive parameters such as V are held constant,

$$F = U - TS \Rightarrow dF = -S\,dt - p\,dV$$

the Gibbs free energy if p, T, and the other extensive parameters are held constant,

$$G = U - TS + pV \Rightarrow dG = -S\,dt + V dt$$

There are four quantities of "thermodynamic potentials" that are useful in the chemical thermodynamics of reactions and non-cyclic processes. The four thermodynamic potentials are internal energy, the enthalpy, the Helmholtz free energy and the Gibbs free energy.

2.8.1. Internal Energy

The energy associated with random and disordered movement of molecules is defines as Internal Energy. There has been a separation that is made in the internal energy from the macroscopic ordered energy associated with moving objects. This refers to the invisible microscopic energy on the atomic and molecular scale.

For Example, a glass of water at room temperature placed on a table has no energy (potential or kinetic). is per second. Now, if the water is tossed across the room, this microscopic energy would not be changed necessarily.

2.8.2. Microscopic Energy

Internal energy is calculated on the microscopic scale. In a given monoatomic gas, the translational kinetic energy of the linear motion of the "hard sphere" type atoms and the Behaviour of the system is defined by the kinetic theory.

However, there is rotational and vibrational kinetic energy for polyatomic gases.

Due to the presence of potential energy in the liquids and solids, there persists intermolecular attractive forces. There must be a simplified visualization of the contributions to internal energy. This is helpful in understanding the phase transitions and other phenomena which involve internal energy.

2.8.3. Helmholtz Free Energy

Helmholtz Free Energy is the thermodynamic potential that helps in measuring the useful work obtainable from a closed system at a constant temperature and volume.

The Helmholtz free energy F is defined by

$$F = U - TS$$

where, F is the Helmholtz Free Energy, U is the internal energy, T represents absolute temperature, and S is the final entropy.

The internal energy U is required to create a system in the absence of changes in temperature or volume. However, if the system is created in an environment having temperature T, then there can be spontaneous energy transfer is TS.

In such cases, there is no need of putting extra energy. It should be noted that if a more disordered final state is created that have higher entropy, then there is less work required to create the system. The Helmholtz free energy is then used as a measure of the amount of energy to be put in to create the system once there is spontaneous energy transfer to the system from the environment.

2.8.4. Enthalpy

Enthalpy is equal to the sum of internal energy and the product of its pressure and volume. Enthalpy is defined by the mathematical formula:

$$H = U + PV$$

where P and V represents pressure and volume, and U is internal energy. Enthalpy is a precisely measurable state variable, because of its explanation in terms of three other precisely definable state variables. Enthalpy is somewhat parallel to the first law of thermodynamics for a constant pressure system.

$Q = \Delta U + P\Delta V$ since in this case $Q=\Delta H$

In order to track the chemical reactions, enthalpy is useful quantity. There are cases in which due to some exothermic reaction, some energy is released to a system. Enthalpy here is useful to show up in some measurable form in terms of the state variables.

An increase in the enthalpy $H = U + PV$ is resulted by an increase in internal energy which could be measured by calorimetry, or with work done by the system, or a combination of the two.

The internal energy U might be the energy required to create a system in the absence of changes in temperature or volume. However, if there is change in the volume, as in a chemical reaction which produces a gaseous product, then emphasis must be given to produce the change in volume.

In case of a constant pressure process, the work should be carried out to produce a volume change ΔV is $P\Delta V$. Then the term PV can be interpreted as the work that must be done to "create room" for the system if it is presumed that it started at zero volume.

2.8.5. System Work

In thermodynamic system, if there is a work done in the system, then it is the gas that is doing the work.

The work done by a gas at constant pressure is:

$$W =_P \Delta_V$$

The process taking place is the area that can be visualized under the pressure-volume curve, in case of a non-constant pressure.

The general expression for work done is represented by:

$$W = \int_{V_1}^{V_2} PdV$$

As per the first law of thermodynamics, the work done by a system decreases the internal energy of the system. System work is an important topic while discussing the heat engines.

2.8.6. Gibbs Free Energy

The Gibbs free energy G is defined by:

$$G = U - TS + PV$$

where, G is the Gibbs free energy; U is the internal energy; T and S are the absolute temperature and final entropy, respectively; P is the absolute pressure; and V is the final volume.

The Gibbs free energy is the thermodynamic potential that can be used to detect the maximum amount of reversible work that is performed by a thermodynamic system at a constant temperature and pressure. As in the case of enthalpy, there must be an additional amount of work PV done in order to "create room" for the system, if the system is created from a very small volume.

In the Helmholtz free energy, there would be a reduction in the overall investment necessary for creating the system, if the work done was at a constant temperature T and it contributes an amount of TS to the system.

The net energy contribution for a system created in environment temperature T with the use of negligible initial volume is the Gibbs free energy.

The change in Gibbs free energy, ΔG, is a very useful parameter in any reaction. Gibbs free energy can be defined as the maximum amount of work obtainable from a reaction.

2.9. CONCLUSION

Thermodynamics conveys the macroscopic descriptions of the complex states of the systems when they interact in an environment. It is limited by the boundaries of a particular nature. The space outside the thermodynamic systems is its surroundings. The Processes that affect the interior of the region of the system are studied explicitly using general principles, these are the four laws of thermodynamics.

The thermodynamics laws were historically established before modern physics. Most of these laws are based on the fact, a system is governed by general statistical principles which apply to the random motion of its particles.

The chapter reviewed the four laws of the thermodynamics. Basically, it discussed the zeroth law of the thermodynamics that stated that the objects are in thermal equilibrium and the net heat exchanged is zero. The first law of thermodynamics revealed that the total energy of the system is the heat gained minus the work done of the system. The second law and the concept of entropy was investigated in the chapter. Entropy is the measure of the disorder of the amount of heat transferred to the system. The third law of thermodynamics stated that the value of the absolute zero cannot be achieved. The potentials of the thermodynamics are also explained which stated the four quantities useful in the chemical thermodynamics. This can be concluded that the overall chapter has explained all about the laws of thermodynamics and its potentials.

REFERENCES

1. Basic Thermodynamics, (n.d.). [ebook] Available at: https://arxiv.org/ftp/arxiv/papers/1501/1501.07088.pdf (accessed on 11 March 2020).
2. Chemistry LibreTexts, (2019). *1st Law of Thermodynamics*. [online] Available at: https://chem.libretexts.org/Bookshelves/Physical_and_Theoretical_Chemistry_Textbook_Maps/Supplemental_Modules_(Physical_and_Theoretical_Chemistry)/Thermodynamics/The_Four_Laws_of_Thermodynamics/First_Law_of_Thermodynamics (accessed on 11 March 2020).
3. Chemistry LibreTexts. (2019). *2nd Law of Thermodynamics*. [online] Available at: https://chem.libretexts.org/Bookshelves/Physical_and_Theoretical_Chemistry_Textbook_Maps/Supplemental_Modules_(Physical_and_Theoretical_Chemistry)/Thermodynamics/The_Four_Laws_of_Thermodynamics/Second_Law_of_Thermodynamics (accessed on 11 March 2020).
4. Ftexploring, (n.d.). *Examples of the Second Law of Thermodynamics*. [online] Available at: https://www.ftexploring.com/energy/2nd_Law-b.html (accessed on 11 March 2020).
5. Ftexploring.com. (n.d.). *Examples of the First Law of Thermodynamics*. [online] Available at: https://www.ftexploring.com/energy/first-law_p2.html (accessed on 11 March 2020).
6. Hyperphysics, (n.d.). *First Law of Thermodynamics*. [online] Available at: http://hyperphysics.phy-astr.gsu.edu/hbase/thermo/firlaw.html#c2 (accessed on 11 March 2020).
7. Hyperphysics, (n.d.). *Internal Energy*. [online] Available at: http://hyperphysics.phy-astr.gsu.edu/hbase/thermo/inteng.html#c2 (accessed on 11 March 2020).
8. Hyperphysics.phy-astr.gsu.edu. (n.d.). *Helmholtz and Gibbs Free Energies*. [online] Available at: http://hyperphysics.phy-astr.gsu.edu/hbase/thermo/helmholtz.html#c2 (accessed on 11 March 2020).
9. Lim, A., (2019). *What is the Zeroth Law of Thermodynamics?*. [online] ThoughtCo. Available at: https://www.thoughtco.com/zeroth-law-of-thermodynamics-4177952 (accessed on 11 March 2020).
10. Lucas, J., (2015). *What is Thermodynamics?*. [online] livescience.com. Available at: https://www.livescience.com/50776-thermodynamics.html (accessed on 11 March 2020).
11. More, H., (2017). *First Law of Thermodynamics and Its Application*

to Different Processes. [online] Redefining Knowledge. Available at: https://hemantmore.org.in/science/chemistry/first-law-thermodynamics/1892/ (accessed on 11 March 2020).

12. Physicsabout, (2019). *Second Law of Thermodynamics Examples and Applications*. [online] Available at: https://physicsabout.com/second-law-of-thermodynamics/ (accessed on 11 March 2020).
13. Review of Fundamental Principles – Thermodynamics: Part I, (n.d.). [ebook] Available at: https://nptel.ac.in/content/storage2/courses/112105129/pdf/RAC%20Lecture%204.pdf (accessed on 11 March 2020).
14. Science ABC, (2017). Why the Universe Will Die a Cold, Dark Death? [online] Available at: https://www.scienceabc.com/nature/universe/what-is-third-law-thermodynamics-definition.html (accessed on 11 March 2020).
15. Thermodynamics Zeroth Law, (2013). [ebook] Available at: https://www.longdom.org/open-access/thermodynamics-zeroth-law-2157–7048.1000e116.pdf (accessed on 11 March 2020).

CHAPTER 3

Entropy and Entropy Components

CONTENTS

Entropy is an important term in the field of thermodynamics and holds a great significance in the measure of various energies that are involved in the processes related to thermodynamics. The measure of entropy provides an indication of the randomness in the system of particles and also gives the details regarding the unused form of energies which cannot be considered for useful work.

This in turn can also provide the energy that is exactly required for a work to be done and the amount that must be supplied in order to achieve that number, thus reducing the consumption and increasing the efficiency of the system. This chapter makes the readers aware about the various terminologies and concepts related to entropy and its possible use in the future.

3.1. INTRODUCTION

Once Von Neumann stated in a famous mention that "Whoever uses the term 'entropy' in a discussion always wins since no one knows what entropy really is, so in a debate one always has the advantage." These words by a great and renowned scientist indicate how entropy has been misunderstood as a concept widely throughout the world.

The understanding of entropy is a concept that lacks concrete evidence to prove that it exists, which makes it even harder to explain it to the people. Whereas, the other thermodynamic properties such as energy, temperature, and pressure can be explained in a much easier way as they can be found evidently across various forms.

Energy as a thing is encountered by us in many walks of life such as the conversion of sunlight to food by photosynthesis in plants or the utilization of energy from the food, we burn in our body so that we can carry out various daily activities and exercises with ease. On the other hand, the encounter with the occurrence of entropy does not take place on such regular basis outside the scope of textbooks and academics.

In 2002, Carson, and Watson found that there is a lot of misconception among the students of the various undergraduate degree programs regarding the subject of entropy and a lot of such misconceptions had grown even before they had taken up the course on physical chemistry.

In 2007, Sozbilir, and Bennet studied a group of undergraduate students who had taken up physical chemistry as the subject for their graduation course and found that a majority of the students were not able to use the thermodynamic principles that they had learnt, to explain the concept of entropy and the change in entropy that took place in a system.In 2014, Kostic reached to a conclusion that the concept of entropy was misused quite often in the literature that was supposed to explain and elaborate on it, even in the field of science. To highlight the significance of entropy in the natural sciences, Sir Edington famously stated that "The law that entropy always increases holds, I think, the supreme position among the laws of Nature…."Entropy has several components such as residual entropy and it gets even harder to explain those components as compared to the concept of entropy. This happens due to the fact that the students need to understand the concept of entropy at the molecular level to understand residual entropy. All this makes it difficult to understand the concept of entropy and grasp the basics in it.

The word Entropy came from the study of heat and energy in the period 1850 to 1900. Some very useful mathematical ideas about probability calculations emerged from the study of entropy. These ideas are now used in Information theory, Chemistry, andother areas of study.

3.1.1. The Clausius Inequality

The first law is simply an energy balance. However, the second law leads to an inequality; an irreversible process is less efficient than a reversible process. Another important inequality in thermodynamics is the Clausius inequality:

$$\oint \frac{\delta Q}{T} \leq 0$$

That is, the cyclic integral of $\delta Q/T$ is always less than or equal to zero. This is valid for all cycles, reversible or irreversible. For internally reversible cycles, it can be shown that:

$$\oint \frac{\delta Q}{T}\bigg|_{int,\,rev} = 0$$

3.1.2. Entropy

The Clausius inequality forms the basis for the definition of a new property called entropy. As can be seen in the equation above, for an internally reversible process the cyclic integral of $\delta Q/T$ is zero. A quantity whose cyclic integral is zero depends on the state only and not the process path, and thus it is a property. Clausius in 1865 realized that he discovered a new property and he called it entropy:

$$dS = \left(\frac{\delta Q}{T}\right)_{int,\,rev} \qquad (KJ / K)$$

Entropy per unit mass is designated by s (kJ/kg. K). The entropy changes of a system during a process can be calculated:

$$\Delta S = S_2 - S_1 = \int_1^2 \left(\frac{\delta Q}{T}\right)_{int,\,rev} \qquad (kJ / K)$$

To perform this integral, one needs to know the relation between Q and T during the process. Note that the cyclic integral of δQ/T will give us the entropy change only if the integration carried out along an internally reversible path between two states. For irreversible processes, we may imagine a reversible process between the two states (initial and final) and calculate the entropy change (since entropy is a property).

3.1.3. The Increase of Entropy Principle

The entropy changes of a system or its surroundings can be negative; but entropy generation cannot.

$$S_{gen} = \begin{cases} > 0 & \text{irreversible process} \\ = 0 & \text{reversible process} \\ < 0 & \text{imposible process} \end{cases}$$

1. A process must proceed on the path that complies with the rise of entropy principle, $S_{gen} > 0$. A process that violates this principle is impossible.
2. Entropy is a non-conserved property, and there is no such thing as the conservation of entropy. Therefore, the entropy of universe is continuously increasing.
3. The performance of engineering systems is degraded by the presence of irreversibility. The entropy generation is a measure of the magnitudes of the irreversibilities present during the process.

3.2. TERMS RELATED TO ENTROPY

3.2.1. Entropy Balance

Entropy is a measure of molecular disorder or randomness of a system, and the second law states that entropy can be created but it cannot be destroyed.

The increase of entropy principle is expressed as

Entropy change = Entropy transfer + Entropy generation

$$\Delta S_{system} = \Delta S_{transfer} + \Delta S_{gen}$$

This principle is commonly known as the entropy balance.

3.2.2. Entropy Change

The entropy balance is easier to apply that energy balance, since unlike energy (which has many forms such as heat and work) entropy has only one form. The entropy change for a system during a process is:

Entropy change = Entropy at final state - Entropy at initial state

$$\Delta S_{system} = \Delta S_{final} - \Delta S_{initial}$$

Therefore, the entropy change of a system is zero if the state of the system does not change during the process. For example, entropy change of steady flow devices such as nozzles, compressors, turbines, pumps, and heat exchangers are zero during steady operation.

3.2.3. Mechanisms of Entropy Transfer

Entropy can be transferred to or from a system in two forms: heat transfer and mass flow. Thus, the entropy transfer for an adiabatic closed system is zero.

Heat Transfer: Heat is a form of energy that is not organized and some disorganization (entropy) will flow with heat. Heat rejection is the only way that the entropy of a fixed mass can be decreased. The ratio of the heat transfer Q/ T (absolute temperature) at a location is called entropy flow or entropy transfer.

Mass Flow: Mass contains entropy as well as energy, both entropy and energy contents of a system are proportional to the mass. When a mass in the amount of m enters or leaves a system, entropy in the amount of ms (s is the specific entropy) accompanies it.

3.3. KINDS OF ENTROPIES

In the field of thermodynamics, entropy is related mainly with three terms:

- Thermal entropy;
- Residual entropy; and
- Total system entropy.

3.3.1. Thermal Entropy

The occurrence of thermal entropy as a property in a system may be attributed to the motion of particles in the system (Boltzmann, 1974). The motions of the particles may be in various forms such as that of rotatory form, vibratory form, translator form or energized form from electronic or nuclear energy. Hence, it can be understood that thermal energy is the sum total of all the energy stored in the system due to the motion of molecules in chaotic ways. The thermal entropy of a system S_{TM} given at a temperature T, may be defined in the form

$$S_{TM} = \Delta_{pt}S + \int_{T}^{\tau} \frac{C_P(T)}{T} dT$$

In this equation C_p may be regarded as the heat capacity of the system at constant pressure, T is the temperature and $\Delta_{pt}S$ is the change in entropy that takes place due to the transitions in phase.

$\Delta_{pt}S$, the entropy of the phase transition, can be calculated by focusing on the enthalpy of phase transition, which is given by $\Delta_{pt}H$ and the temperature at which the phase transition takes place, which is given by T_{pt}.

$$\Delta_{pt}S = \Delta_{pt}H / T_{pt}$$

Theoretically, this equation must be employed in the phase transition of first order as these transitions generally appear as extremely tall peaks that have almost no width. This means that the integral value of C_p/T must be equal to zero.But when the phase transitions are considered and observed in practice, it is found that they do not have the infinite peaks of narrow width. They are found to have finite height and some width to show. Hence, the calculation of $\Delta_{pt}S$ is carried out in four steps. These four steps can be given as:

- The data corresponding to the heat capacity must be fit at the temperatures that lie below the phase transition and above it.
- The heat capacity that has been thus fit, must be extrapolated underneath the peak corresponding to the phase transition.
- The heat capacity thus extrapolated is supposed to be subtracted from the actual heat capacity. After the subtraction process thus performed, the heat capacity obtained is the one that corresponds to the phase transition and is given by $C_{p.\,pt}$.
- After this, the heat capacity of the phase transition has to be fit and integrated by making use of the equation.

$$\Delta_{pt}S = \int (C_{p.pt} / T)\,dT$$

As this term can be seen expressed in the same form as the second term present on the right-side of the equation

$$S_{TM} = \Delta_{pt}S + \int_{T}^{\tau} \frac{C_p(T)}{T} dT$$

This equation may be further simplified in the form:

$$S_{TM} = \int_{T=0}^{\tau} \frac{C_p(T)}{T} dT$$

The amount of heat that is received by the system is very small to an infinite extent and is given as $dQ = C_p$. dT. Due to this, the thermal entropy is associated with the heat that is supplied to the system during the process of heating it from absolute zero temperature to the current temperature T.

As the heat thus generated was stored in the system as the energy due to the motion of the molecules in that system, the thermal entropy is found to be proportional to the heat. Hence, it may be safe to say that the thermal entropy is the kind of entropy that may be found as the result of the motion of the molecules in a system.

Hence, it can also be concluded that the thermal entropy is the indication of the extent to which the various kinds of the kinetic energies stored in the system are found, which cannot be converted to the work that may be useful at a certain temperature (Clausius, 1865, 1867, 1870).

3.3.2. Residual Entropy

The residual entropy, given by S_0, can be found to be the property of a thermodynamic system, which appears near or at the absolute zero. The residual property may be found to be a result of the various kinds of molecular arrangements existing in a random manner. This arrangement is of various asymmetrical particles that align themselves in a crystal lattice. (Clayton and Giauque, 1932; Johari, 2010; Kozliak, 2007; Kozliak and Lambert, 2008; Popovic, 2014, 2015; Sestak et al., 2011).

In 2007, Kozliak stated four ways in which the residual entropy could be calculated. The traditional method to do so is the informational or the combinatorics method which was derived from a model which depended on the tossing of a coin and has been a part of the textbooks for a long time. In this method, the formula used is that of the Boltzmann-Plank one, which is given as:

$$S_0 = k_B \ln \frac{W_{rand}}{W_{perf}}$$

In this equation, k_B is the Boltzmann constant and the number of the microstates of the real, that is the imperfect, crystals, is given by W_{rand} while the number of the perfect crystals are given by W_{perf}.

As given in the definition, W_{perf} is equal to 1. The quantity W_{rand} is connected to the various number of unique kinds of orientations that can be found in a molecule of an imperfect crystal, given by m and the number of molecules that are found in the crystal as a part of its formation, given by N. By definition,

$$W_{rand} = m^N$$

Hence, the residual entropy may be expressed as:

$$S_0 = N.\, k_B.\, \ln m$$

As per this equation, the residual energy is found to be related and dependent on only the number of orientations that can be found to exist among the molecules. Hence, the residual energy may be expected to be dependent only on the number of molecules that are there in the crystal and the arrangement that they make. This residual energy is thus, not dependent on the motion of the molecules and the temperature. Hence, the residual energy is said to be found even at the absolute zero.

Both the kinds of entropies discussed thus far – the thermal and the residual entropy – tend to measure the uncertainty.

However, the kind of uncertainty that they express, or measure may be found to be different.

The residual entropy is the one that measures the uncertainty in the arrangement of the molecules while the thermal energy is the one that measures the uncertainty in the motion of the molecules and the energy that they generate. If the equations of both the entropies are compared with each other, the main point of difference between the residual and the thermal entropy may be easily found. From the equation related to the residual entropy, it can be stated that the residual entropy is connected with the number of molecules in the crystal and the various kinds of unique orientations that they can exhibit (m).Hence, the residual entropy is generated by the uncertainty that can be found to exist in the orientations of the molecules and thus can be said to be a static property.

Whereas, from the equation pertaining to the thermal entropy, it can be stated that the thermal entropy depends on the temperature of the molecules in the system and is generated by the motion of the molecules in it. Hence, the thermal entropy can be said to be the dynamic property of the system which is related to the energy of the system (Clausius, 1865, 1867, 1870).At the absolute zero temperature, the molecules in the system are found to be in the ground state, which shows why there is no uncertainty in the energy that they possess. At this point of time, where the uncertainty in the energy possessed by the molecules cannot be found, the thermal entropy is taken to be zero.

3.3.3. Total System Entropy

The entropy of the system may be considered as an additive property. Hence, it can be stated that the total entropy possessed by the system is the sum of the two types of entropies that have been discussed above, which is the residual and the thermal entropy. Thus, the total entropy of a thermodynamic system S which exists at a temperature T can be calculated from the equation:

$$S = S_0 + \int_{T=0}^{T} \frac{C_p(T)}{T} dT$$

In this equation the residual entropy is given by S_0, the temperature of the system is given by T, while the term $C_p(T)$ denotes the heat capacity of the system at a constant pressure. The term in the integration part of the equation indicates the thermal entropy of the system. Thus, the total system entropy can be considered as the sum of the thermal and the residual entropies of the system. Hence the total entropy of the system may be written as:

$$S = S_0 + S_{TM}$$

In this equation, $S_{TM} = \int (C_p/T)\, dT$ and $S_0 = N\, k_B \ln m$

By this formula of the total system entropy and the equation describing it, the various components of the total entropy of the system can be seen with ease. The total system entropy is the entropy of the system that is indicated in the various data tables pertaining to thermodynamics.

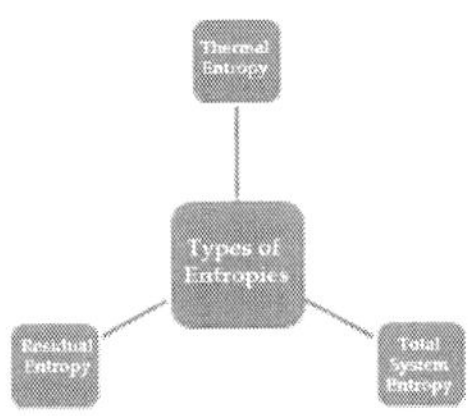

Figure 3.1. Types of entropies.

3.4. ENTROPY AND THE SECOND LAW OF THERMODYNAMICS

Suppose that there is some ice in a drink. It will slowly melt away making the drink cooler. There will be a time when the liquid will reach the thermal equilibrium, which is the same thing as has been predicted by the Second Law of Thermodynamics.

This phenomenon can be expressed in one more way, or the other way to put it is that the Second Law of Thermodynamics can be expressed in one more way. This way of expressing the second law is related to the concept of entropy.

As the various concepts of entropy are explored, it will be seen that the various theories related to the second law of thermodynamics, such as the transfer of heat from hot to cold objects, hold a relation to the natural tendency of the system and the objects to attain uncertainty and become disordered, which makes it difficult for the energy to be available as useful for work.

Thus, the entropy of the system may be expressed as the extent to which the uncertainty in the system exist and the less amount of energy available to do any useful work.

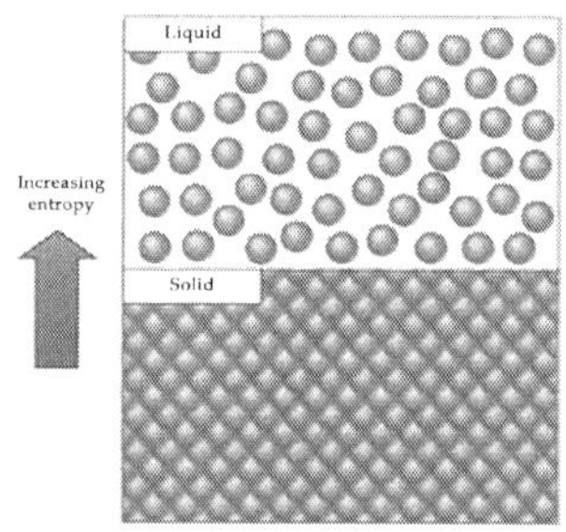

Figure 3.2. *The entropy is more in the liquids than in the solids.*

Source: Image by Wikimedia Commons.

3.4.1. Relating the Entropy, Energy, and Work

The best way to express energy is its simplest definition which is the 'ability to do work.' The entropy may be considered as the measure of energy that is not available to do work. It has been states quite regularly that all the forms of energy can be converted to other forms and used accordingly to do desired work. However, in practice, it is not possible to convert all the forms of energy to the desired form at all the times and utilize it as work.

This energy which is unavailable at the end of the conversion process or in the processes which cannot be carried out, creates interest in the field of thermodynamics as the field concentrates on the conversion of heat to work.

Hence, the entropy may be defined for energy involved in Carnot engine. It has been found that in a Carnot Cycle, or for any reversible process,

$Q_c/Q_h = T_c/T_b$

When the terms are rearranged, they may look like:

$Q_c/T_c = Q_h/T_h$

In this equation Q can be seen as the heat transfer. The transfer of heat (Q) is positive for the heat being transferred into a system and negative for the heat being taken out of the system. T can be seen as the absolute temperature at which the reversible process is taking place. The entropy can be expressed in the SI unit form as joules per Kelvin, given as J/K.

In case the temperature changes take place during the process, then it may be good idea, for the changes in temperature that are extremely small, to get them approximated and take T as the average temperature, which makes it possible for the avoidance of the calculation of the integral values to measure ΔS.

The definition of the change of entropy can be found to be valid only for the reversible processes, for instance, the one used in the Carnot cycle. However, the value of change in entropy can be found even for the real and irreversible processes.

The reason for the calculation of the change in entropy even for the real or irreversible processes is that the entropy S of a system, in the same way as the internal energy U, changes only according to the state of the system and does not depend on the manner in which the system reached that state. The entropy is the property pertaining to the state of the system.

Hence, it can be said that the change in entropy, given by ΔS, is the same for the state 1 and state 2, irrespective of the manner in which the change in state has taken place. All that is required to be done is to find a reversible process or imagine one that transforms from state 1 to state 2 and measure the change ΔS for the said process. This will be the change in entropy for any kind of process that transforms state 1 to state 2.

Now, the change in entropy of a Carnot engine is considered and it is done for the heat reservoirs of the Carnot engine for one complete cycle. The hot reservoir can be seen to have a loss of entropy which can be expressed as

$\Delta S_h = -Q_h/T_h$.

This is due to the fact that the transfer of heat occurs out of it and when the heat is taken out from the system, Q has a negative sign attached with it. On the other hand, the cold reservoir can be considered to have a gain in the entropy which can be indicated by

$\Delta S_c = Q_c/T_c$

This is due to the fact that heat is transferred into the system and the heat taken in by the system has a positive sign attached with it. It can be safely assumed that the reservoirs are large enough and the temperatures of the reservoirs are constant. Hence, the total change in the entropy of the system may be given as:

$\Delta S_{tot} = \Delta S_h + \Delta S_c$

Hence, as it is known that for a Carnot engine

$Q_c/T_c = Q_h/T_h$

$\Delta S_{tot} = (Q_c/T_c) + (-Q_h/T_h)$

Which is equal to zero.

Thus, this result concludes a general validity for the change in entropy of a system which goes on to state that the total change in entropy of a system for any reversible process is equal to zero.

This means that the entropy of the various parts of the system may be different and may change but the overall change in the entropy of the system is zero. In addition to this, the system does not alter the entropy of the environment surrounding it as there is not transfer of heat that takes place between the system and the surrounding.

Hence, it may be stated that the reversible process does not being any change to the total entropy of the system and also does not affect the entropy of the surroundings. At some points of time, this may be stated as follows:

The reversible processes do not affect the total entropy of the universe.

However, the real processes are not reversible in nature and they definitely bring a change to the entropy of the universe. If the value of the entropy of the real and irreversible processes is to be determined, the hypothetical reversible processes may be imagined and used.

Suppose there are two different heat reservoirs, hot, and cold and exchange of heat takes place between them. It is generally found that the entropy increases for the system of two reservoirs that are undergoing this irreversible heat transfer. It can also be established that due to the rise in the entropy of the system, the ability to do work is lost to a great extent. The conclusion is that the entropy of the system is increased, and the energy has become unavailable to do work.If the objects are in the having similar temperatures as of the corresponding object in the irreversible process, the same final state is attained and the same amount of change in entropy is achieved if the heat transfer occurring between the two objects is reversible.

The fact that the entropy of the system increases for any heat transfer that takes place from hot to cold body, is a reasonable one. As the change n entropy is expressed as Q/T, the change in entropy is observed to be quite large at the lower temperatures.

The decrease thus observed in the entropy of the hot object is quite less than the increase in entropy of the cold object, which result in an overall increase in the entropy. This result can be put in a generalized form, which says that:

'*There is an increase in entropy for any system undergoing an irreversible process.*'

As from the point of view of the entropy, two possibilities can be explored:

- The entropy remains constant for a reversible process; and
- The entropy increases for an irreversible process.

The second law, when stated in terms of the entropy of the system, may be put as:

The total entropy of a system either increases or remains constant in any process; it never decreases.

This can be reinstated in the way that the transfer of heat cannot take place from cold body to hot as the entropy of the system would increase. The relation of entropy to energy may be confirmed in the way that the entropy is extremely different from energy.

Entropy may be found to be conserved like energy and increases in all the real processes. The reversible processes, for example, the one in the Carnot engines, may be regarded as the processes in which a lot of heat transfer takes place to work and may also be regarded as the ones in which the entropy remains constant.

3.5. SIGNIFICANCE OF ENTROPY

The thermal property that remains constant in an adiabatic process is called entropy. Alternately, entropy is the physical property of the body that remains constant in an adiabatic process. Entropy is a measure of the part that cannot be used to do work. So, entropy is useful in that it provides the information regarding structural changes accompanying a given process. In a way, entropy is useful for knowing the efficiency of the internal energy of something.

Entropy is very significant in thermodynamics. The significances are mentioned below:

- Entropy is a physical quantity. It is equal to the ratio of heat absorbed or rejected to the temperature;
- It indicates the direction of heat flow;
- It helps in determining the thermodynamic state of an object;
- Like temperature, pressure, volume, internal energy, magnetic behavior it expresses the state of a body;
- The orderliness of an object decreases with the increase of entropy; and
- Like temperature or pressure, it cannot be felt.

Entropy is the measure of the disorder or randomness in a system. In other words, if you drop a box of matches on the floor, the matches will fall all over the place, which is an increase in entropy.

It has been seen that entropy shows an elusive nature. The elusive nature depicted by entropy may be catered to the elusive nature of heat and the motion of the material structure which is related to it. The motion of the material structure is also coupled to the thermal energy through the conversion of the other energy types to heat, in addition to the thermal processes.

The unique form and the distinctiveness of entropy along with the nature of entropy to be universally applicable, derive from the fact that all the processes that take place in the universe in the surroundings at all the points of time and space, happen due to the displacement of mass and energy which has been forced.

This process may be termed as the 'cause and effect' of the manner in which the energy transfer is forced in a certain direction, that is, from higher energy potential to lower energy potential, which is the main force that drive various processes in the universe.

Thus, the various processes that take place spontaneously are due to the forcing of the work and potential energies, which may be coupled with the dissipation of this energy to heat and the related generation of entropy, which the results in the reduction of the work potential, limiting it to nearly zero, at which point the equilibrium is reached.

Typically, with regard to the various processes, that have different material structures which may be functional, man-made or natural in existence and may also include various devices that are functional in nature, the work potential, being an indication of the non-equilibrium, could be transmitted to different forms in numerous ways, in the form of displacement due to forced energy so that various functions and goals could be accomplished. These functions and objectives could also include various processes pertaining to life and its evolution.

The displacement brought about due to the forced energy is mostly coupled with the dissipation of any kind of energy or all of them to the heat or thermal form of energy. This may also include the dissipation of thermal energy to itself from a higher temperature to the lower temperature. This, therefore, makes way for the production of irreversible entropy at all the scales of time and space.

The macro=phenomena related to heat and the properties in relation to it, which also includes the entropy could be put forward as the statistical average of micro particles that are sufficiently large in numbers. It may be noted that at a certain microscale, the thermodynamics becomes the dynamics between the particles.

In various processes that are ideal and reversible, the entropy is considered to be conserved. However, there is no way in which the entropy can be destroyed. This is true even in the cases where the heat is found to be converted to work as the entropy is believed to not be associated with work at all.If the entropy is destroyed, it would mean that there is a transfer of heat spontaneously from lower temperatures to higher temperatures.

It could also mean that there is a heat engine that has an efficiency greater than the ideal and reversible Carnot engine, based on the Carnot cycle, which would mean that it is against the directional and thermal forcing and is not along the lines with the existence of equilibrium in nature.

A micro particle or an atom or molecule, on its own cannot be expected to have a certain temperature or be in possession of a certain amount of heat or thermal energy. IT can also not be expected to have a certain amount of entropy with it.

All these properties are micro properties related to heat, which are found in a larger assembly of micro-particles. These properties may be expected to be there in case the micro particle is a part of the thermal interactions of the large number of particles which create a chaos and an environment of randomness.

However, if observed on a small and individual scale, the entropy related to a micro particle cannot exist and hence is zero. However, within a larger structure, which is a collection of the micro particles that have a thermal randomness among them due to interactions and collisions, the entropy of the system may increase with the various kinds of displacements such as that of time and space and may go on even till the point of infinity.

These parameters catering to heat, characterize the physical state of atoms and molecules in a collective manner, where they may be in any state, be it solid, liquid or gaseous. The various properties such as the temperature, the heat stored, or the thermal energy and the entropy of the system are the macroscopic properties which tend to indicate the thermal and random collective behavior of the various components of a system.

The temperature may be found to represent the average kinetic energy of all the particles that are found in the system. On the other hand, the heat stored in the system or the thermal energy of the system indicates the total amount of the thermal energy present in kinetic and potential form which is possessed by the particles in the system.

In the same way, the entropy of the system may indicate the various displacements or thermal forms that take place within a system and cause chaos in them. This may be termed as the displacement of disordered energy in a random manner, which can be put as the thermal energy that is there without defined structure but is chaotic in nature.

It should be kept in mind that the 'word 'thermal' is an often-overlooked word but is critically important form this point of view. In addition to this, the phenomena pertaining to the molecules are couples and hence may be found to be elusive.

Other than the various thermal interactions and them making way for the rise in the temperature, heat, and entropy of the system, the nature of the various mutual attractions among the atoms and molecules in the various solids and liquids, tend to induce other properties observed at the macroscopic level, which may be termed as the hardness, the elasticity and the tensile strength in the various solids and the viscosity in liquids and other such properties of similar kinds.

Entropy may be considered as a thermal displacement of the thermal energy that takes place due to the temperature acting as a thermal potential ($dQ = TdS$) and thus it can be associated with the thermal heat and also with the absolute temperature. This can be understood as the distribution of the thermal energy among the thermal microparticles that may be moving randomly in space.

The entropy can be considered as the measure of the thermal energy in its random and redistributed form which has been integrated over a certain range of temperature.

The redistribution may have taken due to the transfer of heat or a heat generation which is irreversible in nature and has taken place due to the degradation of energy.

The generation of heat may have more than one aspect to look forward to and may be elusive in nature. This redistribution may take place within a structure of material system in space as per the absolute temperature level and may be given by:

$$dS = dQ_{sys}/T = m\, C_{sys}\, dT/T$$

The unit of this may be J/K.

The entropy of a system may also be given as the measure of the thermal disorder which is given in relation to the logarithm of the number of all the thermal and dynamic microstates, given by W, which is based on the position and the moment of the microparticles and expressed as

$$S = k_B \ln W$$

Some people like to introduce the concept of entropy as the measure of the disorder or the randomness and state that the destruction of entropy is brought about when some order is created, giving the example of the manner in which life evolved.

However, it may be noted that at the time of any of the processes, which may be natural or man-made and which may be of creating or destroying any order or disorder, that is the transformation of any structure of material, the work potential may be regarded as always being dissipated into heat and in such a case the generation of entropy will always be there.

Entropy may be considered to increase from a crystalline structure that is at absolute zero temperature, being uniquely defined, during the process of reversible heating, i.e., the transfer of entropy, or the generation of entropy during the energy conversion in an irreversible process, that is the loss of work potential to the thermal energy that has been produced.

This means that there is degradation of energy or an equi-partition within the structure and space of the material of the system in a random manner, at the level of the absolute zero temperature.

Moreover, the entropy of the system for a given state may be regarded as the same irrespective of whether it is reached at by the heat transfer taking place in a reversible manner or the heat or work dissipation taking place in an irreversible manner.

The various fundamental laws and the physical phenomena may not be guided by certain mathematical models and the results of calculations suggested by some of the equations and theories, but these laws and phenomena may result in the formulation of various theories and equations.

The properties such as the energy, the temperature and especially the entropy of the system are subtle in nature and are hence, elusive, but these properties are well-defined and can be precisely measured in physical forms and used in that way as well. Any new way of approaching the physical phenomena must be based on the knowledge that already exists along with the limitations that are presented with enough clarity and objectivity.

3.6. APPLICATION OF ENTROPY TO THERMODYNAMIC FORMS

Thermodynamic Entropy Applications of thermodynamic principles to the urban system are diverse. A review of some of these is presented by Filchakova and others. For example, in engineering thermodynamics, exergy is defined as the maximum amount of work that may be obtained from a system by bringing it into equilibrium with its environment.

Historically, exergy analysis has been used to improve the thermodynamic efficiency of various industrial processes, by identifying and minimizing exergy destruction. This has led to applications to larger systems assessing energy efficiency at national, regional, and urban scales.

For example, Nielsen, and Jorgensen (2015) developed an exergy accounting framework, mapping the locations of large exergy consumption across six societal sectors for a small region, which allowed for the identification of key areas of attention for a proposed 'sustainable energy' transition plan. It can be shown that the exergy destruction of a process, Ψ_D, is related to its entropy production, S_G, through the environmental reference temperature, T_0, $\Psi_D = T_0 S_G$.

Thus, similar approaches considering entropy as a geographically applied indicator for targeting and improving upon energy inefficiencies are presented in References. These methods remain superior to traditional energy accounting methods, as they capture the 'quality' of energy, that is, its capacity to perform work.

One approach to modern thermodynamics, which departs somewhat from the familiar statistical mechanics interpretation, is an extension of classical thermodynamics outside of equilibrium, and has received attention in its purported applicability to urban systems.

The vast majority of systems observed in nature are open, out of equilibrium, and undergoing irreversible processes. The urban system itself is one such example. Classical thermodynamics, concerned only with the initial and final states of systems in thermodynamic equilibrium, fails to include a theory of irreversible processes.

A thermodynamic study of real systems requires a more general approach, something developed from the beginnings of the 20th century by theorists such as Onsager and Prigogine.

However, this area of thermodynamics is still an active work in progress and lacks an established corpus.

Prigogine, in his study of the phenomena of self-organization, forwards the notion of 'dissipative systems' to refer to complex open structures that maintain their functioning through the constant dissipation of thermodynamic entropy.

This usually involves consideration of the entropy balance representing the total entropy change of the system as the respective sum of its internal entropy production and entropy exchange due to fluxes of matter and energy across the system's boundary:

$$\frac{dS}{dt} = \frac{d_iS}{dt} + \frac{d_eS}{dt}$$

The second law tells us that the internal entropy production $d_iS/dt > 0$, but the total entropy change of the system is permitted to be positive or negative, as $d_eS/dt = d_eS(in)/dt - d_eS(out)/dt$ may take on either sign. Consideration of the city as a dissipative system in relation to a notion of sustainability is considered by Rees and Wackernagel, (1997).

They describe cities as "entropic black holes," which draw in large amounts of entropy and matter, and "export the resultant entropy (waste and disorder)" to maintain their "highly-ordered dissipative structure."

It is argued that the ordered dissipative structure of the city is maintained at the expense of "increasing entropy or disorder in the environment," citing trends of natural capital depletion, pollution emissions, and other adverse anthropogenic ecological consequences.

Marchettini and others. (2006) extend this thread of work, but caution against an "entropic euthanasia"

Where the urban system has fully degraded all potentials through unprecedented growth in energy and matter inflows, leading to "maximum disorder and maximum entropy" analogous to an 'urban heat death.' Similar arguments are also explored.

Filchakova et al. (2007) consider similar arguments in their wider review of thermodynamic concepts applied to the city, exploring its casting as an open system, and the analogy of the city as a living organism. Following the arguments put forward by

Marchettini and others. They appear to strive toward a more quantitative thermodynamic representation to "represent urban metabolism in an operational way."

3.6.1. Interdisciplinary Applications of Entropy

Although the concept of entropy was originally a thermodynamic construct, it has been adapted in other fields of study, including information theory, psychodynamics, thermo-economics/ecological economics, and evolution. For instance, an entropic argument has been recently proposed for explaining the preference of cave spiders in choosing a suitable area for laying their eggs.

3.6.2. The Arrow of Time

Entropy is the only quantity in the physical sciences that seems to imply a particular direction of progress, sometimes called an arrow of time.

As time progresses, the second law of thermodynamics states that the entropy of an isolated system never decreases. Hence, from this perspective, entropy measurement is thought of as a kind of clock.

Figure 3.3. *The entropy study can be used effectively to know about time and related aspects.*

Source: Image by Wikipedia.

3.6.3. Cosmology and Entropy

Since a finite universe is an isolated system, the Second Law of Thermodynamics states that its total entropy is constantly increasing. It has been speculated, since the 19th century, that the universe is fated to a heat death in which all the energy ends up as a homogeneous distribution of thermal energy, so that no more work can be extracted from any source.

Figure 3.4. *Entropy can be used significantly to answer various questions on cosmology.*

Source: Image by Pixabay.

If the universe can be considered to have generally increasing entropy, then – as Sir Roger Penrose has pointed out – gravity plays an important role in the increase because gravity causes dispersed matter to accumulate into stars, which collapse eventually into black holes.

The entropy of a black hole is proportional to the surface area of the black hole's event horizon. Jacob Bekenstein and Stephen Hawking have shown that black holes have the maximum possible entropy of any object of equal size.

This makes them likely end points of all entropy-increasing processes, if they are totally effective matter and energy traps. However, the escape of energy from black holes might be possible due to quantum activity, see Hawking radiation. Hawking has recently changed his stance on some details, in a paper which largely redefined the event horizons of black holes.

3.6.4. Thermodynamic and Statistical Mechanics Concepts

- Entropy unit – a non-S.I. unit of thermodynamic entropy, usually denoted "e. u." and equal to one calorie per Kelvin per mole, or 4.184 Joules per Kelvin per mole.[66]
- Gibbs entropy – the usual statistical mechanical entropy of a thermodynamic system.
- Boltzmann entropy – a type of Gibbs entropy, which neglects internal statistical correlations in the overall particle distribution.
- Tsallis entropy – a generalization of the standard Boltzmann-Gibbs entropy.
- Standard molar entropy – is the entropy content of one mole of substance, under conditions of standard temperature and pressure.
- Residual entropy – the entropy present after a substance is cooled arbitrarily close to absolute zero.
- Entropy of mixing – the change in the entropy when two different chemical substances or components are mixed.
- Loop entropy – is the entropy lost upon bringing together two residues of a polymer within a prescribed distance.
- Conformational entropy – is the entropy associated with the physical arrangement of a polymer chain that assumes a compact or globular state in solution.
- Entropic force – a microscopic force or reaction tendency related to system organization changes, molecular frictional considerations, and statistical variations.
- Free entropy – an entropic thermodynamic potential analogous to the free energy.

- Entropic explosion – an explosion in which the reactants undergo a large change in volume without releasing a large amount of heat.
- Entropy change – a change in entropy dS between two equilibrium states is given by the heat transferred dQ_{rev} divided by the absolute temperature T of the system in this interval.
- Sackur-Tetrode entropy – the entropy of a monatomic classical ideal gas determined via quantum considerations.

CONCLUSION

Total entropy of a thermodynamic system has two components: thermal entropy and residual entropy. Thermal entropy is due to uncertainty in motion of molecules. It is also the measure of useless thermal energy that cannot be converted into useful work. Residual entropy quantifies the uncertainty in molecular orientation. Residual entropy, unlike thermal entropy, is independent of temperature and remains present at absolute zero.

The three kinds of entropy and their relationships can be explained as follows: start by deriving the thermal entropy equation in the usual way, mention that except uncertainty in energy there is uncertainty in orientation as well, resulting in residual entropy, finally show that thermal entropy and residual entropy add up to the total system entropy.

Entropy is applicable across various verticals and field as it physically signifies randomness and its measure. The randomness can be found in any form, be it time or space and hence, various theorists and scientists have pondered over the application of entropy to find the hidden truths not only in the field of thermodynamics but also in the fields of space and time such as that of cosmology.

REFERENCES

1. Bahrami, M., (n.d.). *Entropy*. [ebook] SFU. Available at: https://www.sfu.ca/~mbahrami/ENSC%20388/Notes/Entropy.pdf (accessed on 11 March 2020).
2. Entropy, (n.d.). [ebook] Available at: https://www.basicknowledge101.com/pdf/km/Entropy.pdf (accessed on 11 March 2020).
3. Feistel, R., (2019). Distinguishing between clausius, Boltzmann and Pauling entropies of frozen non-equilibrium states. *Entropy*, [online] *21*(8), 799. Available at: https://www.mdpi.com/1099–4300/21/8/799 (accessed on 11 March 2020).
4. Kostic, M., (2014). The elusive nature of entropy and its physical meaning. *Entropy*, [online] *16*(2), 953–967. Available at: https://pdfs.semanticscholar.org/03c2/b5cf957c9f1746de8e8097bc1168a855449f.pdf (accessed on 11 March 2020).
5. Opentextbc.ca. (n.d.). *Entropy and the Second Law of Thermodynamics: Disorder and the Unavailability of Energy*. [online] Available at: https://opentextbc.ca/physicstestbook2/chapter/entropy-and-the-second-law-of-thermodynamics-disorder-and-the-unavailability-of-energy/ (accessed on 11 March 2020).
6. Popović, M., (2017). Explaining the entropy concept and entropy components. *Journal of Subject Didactics*, [online] *2*(2). Available at: https://www.researchgate.net/publication/324969242_Explaining_the_entropy_concept_and_entropy_components (accessed on 11 March 2020).
7. QS Study, (n.d.). *Significance of Entropy—QS Study*. [online] Available at: https://www.qsstudy.com/physics/significance-of-entropy (accessed on 11 March 2020).
8. Sciencehq.com. (n.d.). Physical significance of Entropy. [online] Available at: https://www.sciencehq.com/chemistry/physical-significance-of-entropy.html (accessed on 11 March 2020).

CHAPTER 4

Thermodynamic Properties of Electrolyte

CONTENTS

The chapter named 'Thermodynamic Properties of Electrolyte' describes the standard thermodynamics properties which are possessed by electrolyte solutions. This chapter mentions the basic concept about the electrolyte and the Q-Electrolattice equation of state (EOS). The topics like electrolyte and non- electrolyte have also been covered in this book. The readers will get insights about the standard thermodynamic properties (STP) possessed by aqueous electrolyte as well as non- electrolytes and also, about the thermodynamic properties that is found in the electrolytes in aqueous solution of glycine.

4.1. INTRODUCTION

Electrolytes are defined as a non-metallic electric conductor in which current is carried by the movement of ions.

The electrolyte solutions can be found in various natural as well as industrial processes. Most of them react with solutions containing dissolved salts in water, for example, wastewater treatment and seawater desalination.

In the case of production of fertilizers, pharmaceutical industries and several processes such as distillation, absorption, and extraction, solutions of non-aqueous electrolytes play a very major role. The prediction and correlation of their thermodynamic properties are quite important for designing and operating these processes.

There are a smaller number of publications regarding the models applied to correlate thermodynamic properties of non-aqueous single salt solutions. The key reason behind this can be the limited availability of experimental data for such systems. The models that are available are based upon the Gibbs excess energy.

Chou and Tanioka in the year 1997 used a modified Debye-Hückel equation to justify and explain long-range interactions along with the concept of solvation to account for short-range interactions. The two-parameter model was applied to correlate vapor pressure of 24 single salt solutions consisting of different kinds of solvents such as ethanol, methanol, acetonitrile, 2-propanol and isopentyl alcohol.

Whereas, Ge et al., (2008) applied the three-characteristic-parameter correlation (TCPC) model to 46 single salts in ethanol, methanol, 2-propanol, acetonitrile, and N-methylformamide, so as to correlate mean ionic activity coefficient, solvent activity and osmotic coefficient.

Additionally, the equations of state (EOSs) have also been applied in determining the thermodynamic properties of electrolyte solutions in non-aqueous solvents. Further, Zuo and Fürst (1997) used an EOS comprised of four terms such as one repulsive term, two attractive short-range terms, and one long-range term, to correlate mean ionic activity coefficients and vapor pressures of single salts in ethanol, methanol, 2-propanol, and acetonitrile solutions.

In the year 2012, Held et al. applied the ePC-SAFT EOS so as to correlate osmotic coefficients, liquid densities and mean ionic activity coefficients at 298.15 K of single salts in ethanol and methanol solutions, offering new experimental data for a lot many electrolyte systems.

In this work, the Q-electrolattice EOS has been successfully applied to aqueous single and mixed salt solutions (Zuber et al., 2014). It is basically used to correlate thermodynamic properties of single salts in ethanol and methanol solutions.

The model comprises the lattice-based Mattedi-Tavares-Castier (MTC) EOS, along with the Born and explicit MSA terms. In this EOS, the cation and anion diameters are fitted, as well as the energetic effects presented by the interactions between solvent, cations, or anions. A model performance analysis is presented depending on the correlations of the thermodynamics properties of sixteen methanol and ten ethanol solutions that contain single strong 1:1 and 2:1 salt.

4.2. ELECTROLYTE

An electrolyte is known as a chemical compound that gets dissociated into ions and as a result, it is able to transport electric charge. In simpler words, an electrolyte is an electric conductor. Unlike several metals, in case of electrolyte the flow of charge is not a flow of electrons, instead it is a movement of ions.

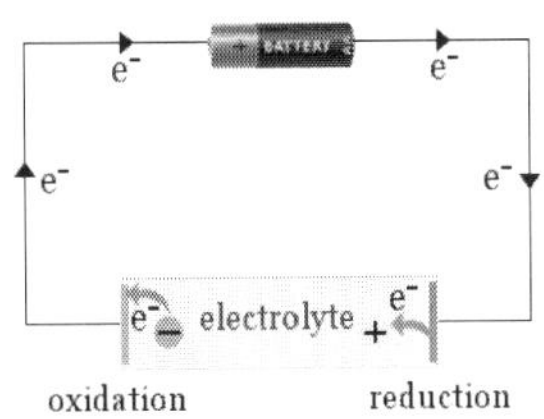

Figure 4.1. *Domestic battery used to attract ions.*

Source: Image by Chemicool.

For instance, the diagram given above shows a domestic battery that is being utilized to attract ions present in electrolyte to the electrodes of an electrochemical cell, where the ions gain electrons or lose electrons. Gaining of electrons is known as reduction, while losing of electrons is known as oxidation.

At times, the electrodes react themselves. For instance, a metal electrode could be oxidized and release ions into the electrolyte solution. Electrolytes can be solid or liquids or solutions. Another definition of electrolytes can be as follows: substances that give ions when they are dissolved in water.

They can be further divided into acids, bases, and salts as they all give ions when they are dissolved in water. The solution conducts electricity because of the mobility of both the positive and negative ions that are known as cations and anions respectively. Strong electrolytes are those that ionize completely when dissolved in water and no neutral molecules are formed in solution.

Example of strong electrolytes:

- NaCl
- HNO_3
- $HClO_3$
- $CaCl_2$

An ionization reaction can be represented as:

$$NaCl_{(s)} \rightarrow Na^{+}_{(aq)} + Cl^{-}_{(aq)}$$

Since, NaCl is an ionic solid (s), which comprises cations Na^+ and anions Cl^- no molecules of NaCl are present in NaCl solid or NaCl solution. The process of ionization is said to be complete. The solute is 100% ionized. Examples of some other ionic solids are as given below:

- $CaCl_2$
- NH_4Cl
- KBr
- CuSO4
- $NaCH_3COO$
- $CaCO_3$
- $NaHCO_3$

Small fractions of weak electrolytes molecules ionize when they are dissolved in water. Some of the neutral molecules are present in their solutions.

For instance, H_2CO_3 (carbonic acid), NH_4OH (ammonia), CH_3COOH (acetic acid), and most organic acids and bases are weak electrolytes.

The following ionization is not complete,

$$H_2CO_{3(aq)} \rightleftarrows H^+_{(aq)} + HCO^-_{3(aq)}$$

In a solution, H_2CO_3 molecules are present. The fraction (generally represented as a%) that undergoes ionization depends upon the concentration of the solution. On the other hand, the process of ionization can be regarded as an equilibrium developed for the reaction given above, for which the equilibrium constant is defined as:

$$K = \frac{[H^+][HCO_3^-]}{[H_2CO_3]}$$

where, [ ] is used to mean the concentration of the species in the [ ]. For carbonic acid, $K = 4.2x10^{-7}$.The ionization or auto-ionization of pure water can be represented by the ionization equation given below:

$$H_2O \rightleftharpoons H^+ + OH^-$$

And, the equilibrium constant is given as:

$$K = \frac{[H^+][OH^-]}{[H_2O]}$$

For pure water, $[H_2O]$ is a constant (1000/18 = 55.6 M), and the ion product, K_w, can often be used for water,

$$K_w = K[H_2O][H^+][OH^-]$$

The constant K_w depends on temperature. At 298 K, $K_w = 1x10^{-14}$.

If no solute is present in water, the solution has equal concentrations of $[H^+]$ and $[OH^-]$.

$$[H^+] = [OH^-] = 1 \times 10^{-7}$$

And,

$$pH = -\log[H^+] = 7$$

It is important to note that, only at 298 K, the pH of water is equal to 7. At much higher temperatures, the pH is quite less than 7. While, at lower temperatures, the pH is greater than 7.

4.2.1. Electrolytes in Body Fluids

Body fluids are solutions of electrolytes and several other things. The combination of blood together with the circulatory system is the river of life as it coordinates all the life functions.

When the heat stops pumping during a heart attack, the life ends quickly. Getting the heart to pump again as early as the patient is critical so as to save his or her life.

The primary electrolytes that are needed in the body fluid are cations (of sodium (Na^+), calcium, magnesium (Mg^{2+}), and potassium) and anions (of carbonates, chloride, phosphates, amino-acetates, and iodide). These are known as micro minerals in context to nutrients.

Balancing of electrolyte is important for several functions taking place inside the body. Given below are certain examples showing what exactly can happen when there is an imbalance of electrolytes:

1. Increased levels of potassium (K^+) may result in cardiac arrhythmias;
2. Decreased extracellular K^+causes paralysis;
3. Too much extracellular Na^+ causes fluid retention; and
4. Decreased plasma calcium (Ca^{2+}) and Mg^{2+} can cause muscle spasms of the extremities.

When a patient is dehydrated, a commercially available electrolyte solution is needed to maintain his health and well-being. When it is about a child's health, oral electrolyte is required when a child suffers from diarrhea.

The use of oral electrolyte solutions that are highly responsible for saving a huge number of people all around the globe over the past two decades, is one of the most crucial medical advances in taking care of health of children in the past one century (Juilus G. K. Goepp, MD, Assistant Director of the Pediatric Emergency Department of the Children's Center at Johns Hopkins Hospital).

If a parent gives an oral electrolyte solution at very initial stage of illness, then dehydration can be easily prevented. The functionality of an electrolyte solution is basically related to its properties.

4.2.2. Electrolytes and Batteries

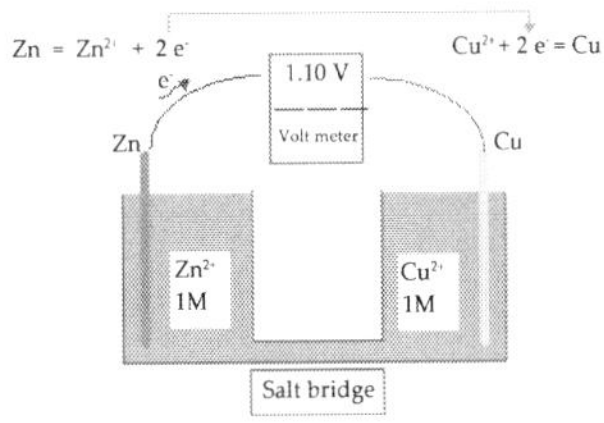

Figure 4.2. *Copper-zinc battery.*

Source: Image by Chemicool.

Electrolyte solutions are needed in batteries always, even in dry cells also. The simplest form of battery comprises two electrodes. The figure given above shows a copper-zinc battery. The electrode present in left is a zinc electrode. The zinc atoms have a tendency to become ions and leave the electrons behind.

$$Zn_{(s)} \rightarrow Zn^{2+}_{(aq)} + 2e^-$$

As the Zn ions starts getting into the solution, anions will move from the copper cell to the zinc cell in order to compensate for the charge. Simultaneously, electrons move from the Zn electrode to the Cu electrode to neutralize the copper ions.

$$Cu^{2+}_{(aq)} + 2e^- \rightarrow Cu_{(s)}$$

The solution in dry cells is replaced by a paste so that the solution will not leak out of the package. In this cell, the Zn and Cu electrode has a voltage of 1.10 V, if the concentrations of the ions are as indicated.

4.2.2.1. Chemical Reactions of Electrolytes

When electrolyte solutions are combined, the cations and anions collide. When the ions are different, then there will be no reaction. Although, certain cations and anions may create a molecule or solid and hence, the cations and anions change their partners. These are thus known as metathesis reactions, which include:

a) **Solid formationor precipitation reactions**: In this reaction, the cations and anions form a less soluble solid. This eventually results in the appearance of a precipitate.

b) **Neutralization reactions:** H^+ of an acid and OH^- of a base combine to give the neutral water molecule.

c) **Gas formationreactions:** When neutral gaseous molecules are formed in a given reaction, they leave the solution forming a gas.

In addition, the redox reactions are also possible between different ions. Actually, the battery operations comprise redox reactions.

4.2.3. Examples of Electrolytes

1. **Molten Sodium Chloride**

Molten sodium chloride (NaCl) acts as an electrolyte in the process of producing Na^+ metal. In the reaction, chlorine gas is also produced. In this case, the electrolyte is consumed by the redox reactions taking place inside the cells.

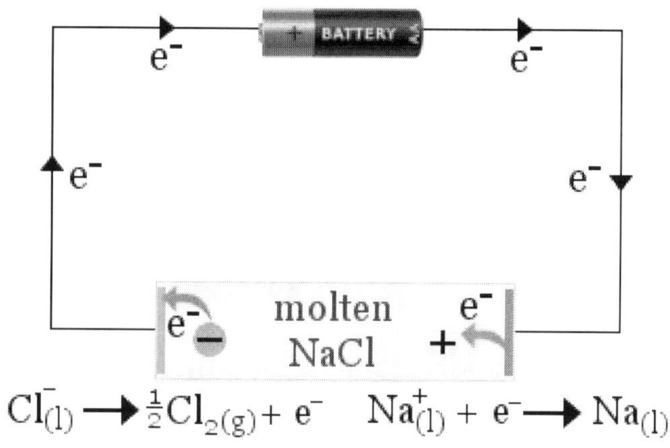

Figure 4.3. *Using Molten NaCl for producing sodium metal.*

Source: Image by Chemicool.

When potassium hydroxide is dissolved in water, it produces a very conductive electrolyte in the Edison cell, an early rechargeable cell.

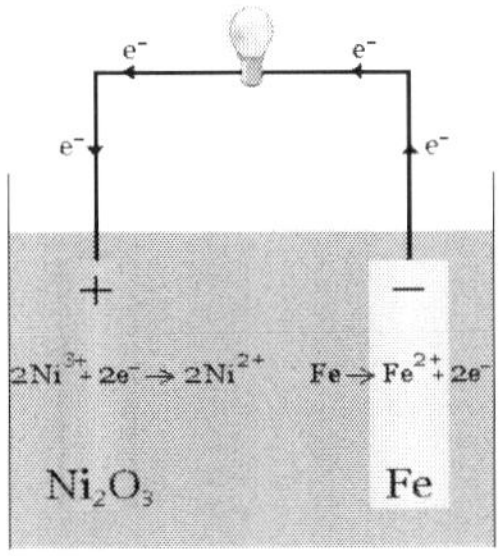

Figure 4.4. *Crude schematic of Edison cell discharging.*

Source: Image by Chemicool.

Different from the case of production of Na^+ given above, no electrolyte is consumed in this reaction. In this case, the role of K^+ hydroxide is to increase the presence of ions in order to increase the conductivity of electrolyte and hence, complete the electric circuit.

The chemical equations for the reactions at the electrodes are:

$$Ni_2O_3 + H_2O + 2e^- \rightleftharpoons 2NiO + 2OH^-$$

$$Fe + 2OH^- \rightleftharpoons Fe(OH)_2 + 2e^-$$

For discharge the equations are read from left to right; for charging right to left.

Electrolytes were liquids traditionally, that allows the movement of ions. Nevertheless, solid materials are now available that allows ion to move in very easy way through their structures. Solid electrolytes are desirable in various consumer products as they do not leak.

Lithium tin phosphorous sulfide, $Li_{10}SnP_2S_{12}$, is a solid electrolyte for use in lithium ion batteries. The crystal structure of the solid electrolyte united with its weak interaction with lithium ions creates an environment that lithium ions can easily hop through.

Electrolytes are very important for the electrochemical processes in living beings. The major ions in these electrolytes are calcium (Ca^{2+}), sodium (Na^+), magnesium (Mg^{2+}), potassium (K^+), hydrogen carbonate (HCO_3^-), chloride (Cl^-) and hydrogen phosphate (HPO_4^{2-}).

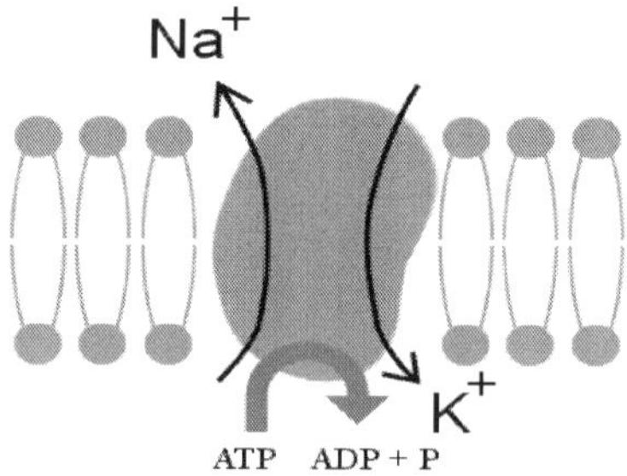

Figure 4.5. *Sodium and potassium pump.*

Source: Image by Chemicool.

The significance of the sodium-potassium pump to our lives can be explained by the fact that it utilizes almost one-third of resting energy. The pump maintains the electrolyte balance of cells, with excess K^+ ions present inside of the cells and excess sodium ions present outside of the cells.

Such concentration gradient develops a voltage across the cell wall that allows electrical signals to get transmitted in muscles and neurons. Also, it provides the energy for certain processes that take place in cell membranes.

4.3. Q-ELECTROLATTICE EQUATION OF STATE

There was a methodology which was proposed by Myers (2002) and this methodology was considered while developing the Q-electrolattice EOS. For that purpose, the interactions between species in an electrolyte mixture are considered and this is done by adding the contribution of different terms to the Helmholtz energy (A^R). This sum is the result of a thermodynamic path which is relevant to describe the behavior of electrolyte solutions accurately. The residual A^Rin order to form an electrolyte solution along the mentioned path is:

$$A^R(T,V,n) = \Delta A^{MTC} + \Delta A^{Born} + \Delta A^{MSA}$$

In this given equation, the ΔA^{MTC} term is the contribution to the A^R related to the repulsive and attractive effects which was originally proposed by Mattedi (1998); the term ΔA^{Born} represents the contribution to the short-range interactions which is related with the solvation effects and the term ΔA^{MSA} represents the long-range interactions because of the presence of electrolytes in the solution (Blum, 1988; Clarke and Bishnoi, 2004; Baseri and Lotfallahi, 2011).

The evaluation of the dielectric constant can be done by coupling the Pottel model for dielectric constant presented by Zuo and Fürst (1997) to the EOS. The change in the dielectric constant of the solvent is considered by this model because of the presence of electrolytes. The dielectric constant of the solvent is expressed in the form of a polynomial equation, (for temperatures in Kelvin), and the equation for the same is mentioned below:

$$D_s = d_1 + \frac{d_2}{T} + d_3 T + d_4 T^2 + d_5 T^3$$

In the equation given above, D_S is the pure solvent dielectric constant, and d_1- d_5 are parameters.

There are certain parameters for the equation provided above and they are mentioned in the table provided below. In the case of ethanol, the parameters are equal to those, which are presented by Zuo and Fürst (1997). In the case of methanol, new parameters have been obtained on the basis of the experimental dielectric constant, which is presented by Davidson (1957) and Shirke et al., (2000), in a temperature range from 176.60 K to 318.15 K. The average absolute relative deviation (AARD (%)) for dielectric constant, in the case of methanol, is obtained with the new set of parameters and that is 1.81%.

Table 4.1. Parameters of Dielectric Constant in Eq. (2)

Solvent	d_1	d_2	d_3	d_4	d_5	**T (K)**
Methanol	1.0462 x 10^2	1.0 x 10^3	9.0108 x 10^{-2}	–2.5998 x 10^{-3}	4.8503 x 10^{-6}	176.60 – 318.15
Ethanol	1.7572 x 10^2	–3.0699	–3.5350 x 10^{-1}	–2.0285 x 10^{-3}	5.0644 x 10^{-6}	288.15 – 328.15

Source: Table by Scielo.br.

4.4. ELECTROLYTES AND NONELECTROLYTES

In broader terms, the classification of the liquid mixtures can be done as either solutions of electrolytes or solutions of non-electrolytes. Electrolytes are considered to be the substance which can dissociate into electrically charged particles known as ions, while non- electrolytes contain the molecules that bear no net electric charge.

Therefore, when the sodium chloride (NaCl) which is the ordinary salt is dissolved in water, it further forms an electrolytic solution. Simultaneously, it dissociates into positive sodium ions (Na^+) and negative chloride ions (Cl^-), whereas on the other hand, when the sugar is dissolved in water maintains its molecular integrity and does not dissociate. Water is the most common solvent for electrolytes because of its omnipresence; the ocean is a solution of electrolytes.

However, the electrolyte solutions are also formed by other solvents (like ammonia and sulfur dioxide) which have more dielectric constant (which can be defined as a measure of the ability of a fluid to decrease the forces of attraction and repulsion between the charged particles).

The energy which is required to separate an ion pair (which is the pair of one ion being positive charge and other ion being negative charge) varies inversely along with the dielectric constant. Therefore, only in the solvents, appreciable dissociation into separate ions occurs with large dielectric constants.

There are some electrolytes such as salts, which are non-volatile. This means that they have essentially no tendency to enter the vapor phase. However, there are some notable exceptions, like hydrogen chloride (HCl), which is readily soluble in water. After being soluble in water, it forms hydrogen ions (H^+) and chloride ions (Cl^-). At normal temperature and pressure, pure HCl is a gas and when there is no water or some other ionizing solvent, HCl exists in molecular, instead of ionic form.

The solutions of electrolyte readily conduct electricity, on the other hand, the non-electrolyte solutions do not conduct electricity. An example for good electrical conductor can be a dilute solution of HCl in water, but a dilute solution of HCl in a hydrocarbon is a good insulator. Because of the fact that there is a large difference in dielectric constants, HCl is ionized in water but not in hydrocarbons.

4.5. STANDARD THERMODYNAMIC PROPERTIES OF AQUEOUS ELECTROLYTES AND NON-ELECTROLYTES

Thermodynamic modeling (TM) is quite important for understanding and predicting phase and chemical equilibria in natural and aqueous systems at high temperatures and pressures. These systems comprise a great variety of organic as well as inorganic solutes ranging from a polar non-electrolyte to strong electrolytes; temperature and pressure affect speciation of solutes that come across in ionic or molecular forms, or as ion pairs or complexes.

Properties in context to the Gibbs energy like thermodynamic equilibrium constants of hydrothermal reactions and activity coefficients of aqueous species, are needed for practical use by geologists, process engineers and power-cycle chemists.

Derivative properties such as heat capacity, enthalpy, and volume can be obtained from volumetric and calorimetric experiments. These properties are very useful in extrapolations while calculating the Gibbs energy at conditions remote from ambient.

They sensitively show evolution in molecular interactions with changing temperature as well as pressure. In this case, models with a key theoretical basis are necessary that can efficiently describe all the thermodynamic functions of an aqueous system over a wide range of temperature and pressure, describing with a limited number of adjustable parameters.

In case of thermodynamics of hydrothermal solutions, the unsymmetrical standard-state convention is mainly used. The standard thermodynamic properties (STP) of a solute shows its interaction with the solvent (that is, water), and the excess properties, correlated to the activity coefficients, correspond to solute-solute interactions.

For dilute and quite concentrated solutions, the standard-state functions have a major role and can be effectively utilized as a reasonable approximation for semi-quantitative modeling. The interactions between solute and solvent prevail at around critical conditions, where all the STP of solutes undergo various changes or modifications. The standard derivative properties of a solute scale with the isothermal compressibility and thermal expansivity of the solvent and diverge at the critical point of the solvent.

The direction of such divergence cannot be clearly predicted for some types of solutes without experimental evidences. In addition to that, this behavior highly affects the properties of solutions in a broad variety of conditions below and above the critical point.

As a result, the modeling approaches used as well as the experimental data available at ambient conditions are not enough for extrapolation of STP towards high temperatures as well as pressures, especially up to the supercritical region, though these modeling approaches are important in calculations.

4.6. THERMODYNAMIC MODELS FOR ELECTROLYTE SOLUTIONS

The development of the thermodynamic models for electrolyte solutions was done in order to be able to mathematically describe the properties and the phase behavior of solutions. If the chemical industries are considered, then it can be said that it is very important to be able to optimize processes for the production of chemicals. Electrolyte solutions are involved in a number of different processes and therefore, it is important to have good models for the description of electrolyte properties.

All different types of interactions such as ion-ion, ion-dipole, dipole-dipole, molecule-molecule should be considered while properly modeling the electrolyte systems. The potential energy which is caused by ion-ion interactions is proportional to the inverse separation distance, 1/r. Therefore, the electrostatic ion-ion interactions have an effect over a relatively long distance and so, are called long-range interactions.

The potential energy which is caused by the molecule–molecule interactions is proportional to the sixth power of the inverse separation distance $1/r^6$. This is reason why these interactions are referred to as short-range interactions.

The potential energy of ion–dipole interactions is proportional to $1/r^2$ and the potential energy of dipole–dipole interactions is proportional to $1/r^3$. So, these interactions can be referred to as intermediate range interactions. A number of models are structured along with terms representing only long range and intermediate/short range interactions.

4.6.1. Electrostatic Interactions

4.6.1.1. Debye–Hückel theory

The very first model which was really successful for the electrostatic interactions between ions in aqueous electrolyte was developed in the year 1923, by P. Debye and E Hückel. They described the thermodynamics of ideal solutions of charged ions. The electrostatic interactions between charged ions only represent the long-range interactions in such solutions and not the short-range interactions, which have already been mentioned above.

The Debye-Hückel model does not described the interactions between ions and water, which has resulted in people to refer this model as a 'dielectric continuum model.' In this model, the solvent only plays a role because of its relative permittivity (dielectric constant) and its density. Therefore, the model provided by Debye Hückel cannot stand alone as a model for electrolyte solutions.

It tends to only represent some of the electrostatic interactions and so, should be combined with a term for short as well as intermediate range interactions in order to wholly describe the properties of concentrated electrolyte solutions. In the Debye-Hückel theory, the electrostatic force which is exerted by a positive ion on a negative through the solvent medium is expressed by using Coulomb's law:

$$F = -\frac{1}{4\pi\varepsilon_0\varepsilon_r}\frac{e^2}{r^2}$$

where, e is the electronic charge = 1.60206 x 10^{-19} C; e_0 is the permittivity in vacuum = 8.8542 x 10^{-12} C^2 J^{-1} m^{-1}; e_r is the dielectric constant (relative permittivity) of the solvent (which is unitless). The value of the relative permittivity of water is 78.4 at temperature of 298.15 K. Here, r is the distance between the ions (in meter).

The equation given by Poisson provides a relationship between the charge density (ρ_i Cm^{-3}) around ion i and the electrical potential (ψ_i J/C) for a sphere having radius r around ion i:

$$\frac{1}{r^2}\frac{d}{dr}\left(r^2\frac{d\psi_1}{dr}\right) = -\frac{\rho_i}{\varepsilon_0\varepsilon_r}$$

Because of the charges, the ions are not distributed evenly or randomly in the solution. When near a cation, anions tend to be in excess whereas, near an anion, cations tend to be in excess. If an ion j is at the distance of r from the ion i, then j will have an electrical potential energy of $z_j e\psi_i$. According to the assumptions done by Debye and Hückel, the distribution of the ions in the solution to be a Boltzmann distribution. This assumption provides another relation in between the charge density and the electrical potential:

$$\rho_i = eN_A \sum_{\text{all ions}} \frac{n_j z_j}{nV} e^{-\frac{z_j e\psi_1}{kT}}$$

where, n_j is the mol number of component j, z_j is the charge of component j, N_A is Avogadro's number which is equal to 6.023 x 10^{23} mol^{-1}, k is the Boltzmann's constant which is 1.381 x 10^{-23} JK^{-1}, and T is the temperature in Kelvin and V is the molar volume of the solution.

The Poisson equation and the Boltzmann equation were combined by Debye and Hückel while eliminating the charge density. The result was the Poisson-Boltzmann equation which was solved for the electrical potential ψ_i.

Debye and Hückel finally arrived at an excess Helmholtz function considering an ideal solution of charged ions in their paper. It was like a contradiction to have an excess Helmholtz function for an ideal solution having charged ions. Ideal solutions do not contain excess terms.

But as it is previously mentioned, this excess Helmholtz function only takes the non-ideality which is caused by the electrostatic interactions into account and it does not deal with the traditional non-ideality, which is caused by short range forces. There is an equation which can express the molar excess Helmholtz function for the electrostatic interactions which can be given as:

$$\frac{A^E}{RT} = -\frac{1}{3}\sum x_i z_i^2 sk\chi(ka_i)$$

The term *s* in the given equation can be defined by:

$$s = \frac{e^2}{4\pi\varepsilon_0\varepsilon_r kT}$$

There was a symbol a_i (a for "annäherungsabstand"), which represents the distance of the closest approach to the ion *i*. The distance of closest approach is said to be a parameter for the radius of ion *i* and not its diameter. It is expected that a_i is larger in comparison to the radius of the ion. The reason is that the ions are thought to be surrounded by water which gives the ions a larger radius than the bare ion.

The product that is κa_i is dimensionless and therefore, κ is a reciprocal length.

1/κ is a characteristic length known as the screening length. The screening length provides a good first estimation about the distance beyond which Coulomb interactions can be essentially ignored, and also, the size of the region near a point charge where opposite-charge counter – ions can be found. The expression for κ can be given as:

$$k = \left(\frac{e^2 N_A}{\varepsilon_0\varepsilon_r kT}\frac{\sum n_i z_i^2}{nV}\right)^{1/2}$$

The function χ can be given as:

$$\chi(x) = \frac{3}{x^3}\left(\frac{3}{2} + \ln(1+x) - 2(1+3) + \frac{1}{2}(1+x)^2\right)$$

The Helmholtz function of Debye and Hückel has not been used by those who have developed models for electrolyte solutions.

4.6.1.2. The Born Equation

The Debye-Hückel theory is associated with the interaction between charged ions, Born derived an equation for the interaction between an ion and the surrounding solvent. There is an electrostatic field around an electrical charge. Such fields affect the polar molecules and redirect themselves to have as low energy as possible in the field.

The positive part of a polar molecule is attracted to a negative charge and the negative part is attracted to a positive charge. Then it can be said that an orientation polarization is taking place. A solvent which contain polar molecules is polarizable and so, is referred to as a dielectric medium. And there are highly polarizable solvents which have high relative permittivity (dielectric constants).

The electrostatic interactions are then lowered by solvents having high relative permittivity. In a vacuum, salts do not dissociate spontaneously, and the reason is that there is an electrostatic interaction between the ions and they are too strong. In water, the electrostatic interactions between ions are lowered by a factor which is represented as $\varepsilon_r = 78.4$ at 25°C.

The water molecules tend to shield the ions from each other and allow them to be separate. The relative permittivity of a solvent can be defined as the ratio in between the permittivity of the solvent and the permittivity of vacuum and so can be represented as:

$$\varepsilon_r \equiv \frac{\varepsilon}{\varepsilon_0}$$

Previously, the relative permittivity was referred to as the dielectric constant and it was represented as D. The measurement of the relative permittivity of a medium can be done by exposing the medium to a perturbing external electromagnetic field of small field strength.

The medium is then placed in between the plates (that is conductors) of a capacitor. Then, the response of the medium to this field is measured. Using this response, the measurement of the relative permittivity of the medium can be done.

When the ions get dissolved in a dielectric medium, then the solvent molecules are polarized by the electrical charges. This interaction between ion-solvent is known as solvation. Considering the special case of water as solvent, the term hydration is used.

The hydration is not the formation of ion-hydrates of any particular stoichiometric composition. It is the redirection of the polar water molecules around the charged ions. The energy change related with the solvation is referred to as the solvation energy or in the case of water, the energy change is referred to as the hydration energy.

Let's consider a spherical ion having charge Z_i and radius r_i in a medium with the permittivity $\varepsilon_0\varepsilon_r$. As per the calculations done by Born, the solvation energy of this ion can be calculated by integrating the energy of the electrical field from the surface of the spherical ion to infinity. Born found that the electrostatic contribution to the A^R of the ion can be given by:

$$A = \frac{Z_i^2 e^2}{8\pi\varepsilon_0\varepsilon_r r_i}$$

This expression was derived for a single ion in a solvent having the relative permittivity ε_r. When there is presence of more ions, then the electrostatic fields, which are generated by the ions influence each other, and these electrostatic interactions influence the interactions between ion-solvent.

The equation which is given above shows that the A^R is lower in a medium of high relative permittivity, ε_r. Therefore, if the ion is transferred from vacuum to a solvent then the solvation energy is negative. By multiplying the energy given in the above equation with Avogadro's number, the A^R of one mol of ions with charge Z_i can be calculated.

Mostly, it is of interest to know the change in A^R when there is movement of one mole of an ion from a medium with one permittivity to a medium with another permittivity. This is referred to as the A^R of transfer. If one mole of ions is moved from vacuum to a medium of relative permittivity ε_r, then the energy change can be given as:

$$\Delta_{solv} A = \frac{Z_i^2 e^2 N_A}{8\pi\varepsilon_0\varepsilon_r r_i} - \frac{Z_i^2 e^2 N_A}{8\pi\varepsilon_0 r_i} = \frac{Z_i^2 e^2 N_A}{8\pi\varepsilon_0 r_i}\left(\frac{1}{\varepsilon_r} - 1\right)$$

It has been found that equation that is given above, provides very accurate results in comparison to the experimental values of the solvation energy. An effective radius corresponding to the radius of the spherical cavity in the solvent, which is created by an ion, should be used in the equation given above instead of the ionic radius of the ion.

As per the views of Rashin and Honig, the radius of the cavity, which is produced by the same ion, differ depending upon different solvents. When the calculation of the A^R of transfer between two solvents is done, then it becomes necessary to use two different radii for the same ion.

On the other hand, Rashin, and Honig found that the calculation of the cavity radius of anions in water can be done by adding 7% to their ionic radius, whereas the calculation of the cavity radius of cations can be done by adding 7% to their covalent radius.

The covalent radius is said to be half the distance between the two identical atoms which are bonded together by a single covalent bond. Usually, the covalent radius of a metal cation is larger in comparison to the ionic radius but smaller in comparison to the atomic radius of the corresponding metal.

The atomic radius of Na^+ is 190 pm, whereas the covalent radius of Na^+ is 154 pm and also, the ionic radius of the sodium ion is 102 pm. The argument for using the covalent radius of the metal instead of the ionic radius of the cation is that the covalent radius corresponds to the radius of the empty orbital around the cation. The assumption is that this empty orbital forms a part of the cavity.

The A^R of transfer is according to Rashin and Honig for the transfer of one mole of ion i from a medium with relative permittivity $\varepsilon_{r,1}$ to a medium with relative permittivity $\varepsilon_{r,2}$ which is given by:

$$\Delta_{\text{transfer}} A = \frac{Z_i^2 e^2 N_A}{8\pi\varepsilon_0 r_{i,2}}\left(\frac{1}{\varepsilon_{r,2}} - 1\right) - \frac{Z_i^2 e^2 N_A}{8\pi\varepsilon_0 r_{i,1}}\left(\frac{1}{\varepsilon_{r,1}} - 1\right)$$

$$= \frac{Z_i^2 e^2 N_A}{8\pi\varepsilon_0}\left(\frac{1}{r_{i,2}\varepsilon_{r,2}} - \frac{1}{r_{i,2}} - \frac{1}{r_{i,1}\varepsilon_{r,1}} + \frac{1}{r_{i,1}}\right)$$

For the two solvents, the radii of the solvent cavities are marked with subscript 1 and 2.

The equation which is provided above has the Born term and this equation is often used with equations of state for electrolyte solutions. In such case, the Born term contributes to the activity coefficients because of the variation of the relative permittivity with pressure. The Born term is also used in order to calculate the Gibbs energy of transfer.

This is the difference that can be seen in the standard state chemical potential for a solute in two solvents having different relative permittivity. Usually, it is assumed that the A^R difference calculated, by using the above equation, is identical to the corresponding difference in Gibbs energy.

The A^R, which is calculated from the equation, is the difference in A^R of a charged particle in a dielectric medium in comparison to the A^R of an uncharged particle of the same size, in the same medium. Thus, the equation represents the contribution to the excess A^R from the interaction that takes place in between a single ion and a number of solvent molecules.

Helgeson used the equation of Born along with the extended Debye-Hückel law in order to derive an activity coefficient model. This activity coefficient model seemed to be identical to the equation derived by Hückel in the year 1925, but it was derived from a number of different theoretical assumptions. Thus, the Helgeson-Kirkham-Flowers model can be seen as a more modern validation of the equation of Hückel.

4.7. THERMODYNAMICS OF ELECTROLYTE SOLUTIONS

4.7.1. Gibbs Energy Function for Real Electrolyte Solutions

The real electrolyte solutions can be thermodynamically defined in terms of their deviation from ideality that is the excess properties.

Let us consider n_s mol of a salt with the stoichiometric coefficients v_i. Then assume that the salt is entirely dissociated into ions in an aqueous solution which contains moles of water which is represented by n_w. Then the Gibbs free energy of the solution can be expressed as

$$G = \sum_i n_i \mu_i = n_w \mu_w + n_s \sum v_i \mu_i \qquad (1)$$

In the above equation, μ_i represents the chemical properties of component *i*.

From classical, there is non–electrolyte thermodynamics which is at constant temperature

$$d\mu_i \equiv RTd(\ln \hat{f}_i) \Rightarrow \mu_i = \mu_i^\bullet + RT\ln\frac{\hat{f}_i}{f_i^\bullet} \qquad (2)$$

In the given equation, R is the gas constant, T represents the absolute temperature, $\hat{f}_i$ represents the fugacity of component *i* in the solution at P and T, $\mu_i^\bullet$ represnts the checmical potential of component *i* in its standard state, and $f_i^\bullet$ represnts the fugacity of the component *i* in its standard state.

There are several different concentration scales which are being used in the elctrolyte thermodynamics, which molality (mol (kg $H_2O)^{-1}$) and mole fraction. The molarity (mol dm^{-3}) concentration scale is dependent on temperature and therefore, it is not practical in a thermodynamic model.

There is variation in the activity coefficients according to the different concentration scales. Activity coefficients on the basis of the mole fraction are referred to as the rational activity coefficients, activity coefficients on the basis of the molality scale are referred to as the practical or molal activity coefficients.

There are two different standard state conventions which are used for the electrolyte solutions.

As per the symmetrical convention, the activity coefficient of a component is equals to one in the pure component (hypothetical) at all the temperatures. As per the unsymmetrical convention, the activity coefficient of a component is equals to one at the infinite dilution at all the temperatures.

The molal activity coefficients are always used along with the unsymmetrical convention. The activity of any component is equals to one in the standard state of the component. Mole fraction along with the symmetrical convention, the fugacity of component i can be expressed as $\hat{f}_i = x_i f_i f_i^{\bullet}$, in which f_i represents the rationa, symmetrical activity coefficient for component i. The chemical potential of component i can be represented by using the equations as:

$$\hat{f}_i = x f_i f_i^{\bullet} \qquad (3)$$

In the equation given above, μ_i^o is the chemical potential of pure component i. Usually in the symmetrical convention, the mole fraction scale is always chosen for the solvent. In the case of unsymmetrical convention, mole fraction scale, the fugacity of component i can be represented as, $\hat{f}_i = x f_i^{\Delta} H_{i,x}$ in which $H_{i,x}$ represents the rational Henry's law constant and f_i^{Δ} is the rational, unsymmetrical activity coefficient for the component i. In the standard state of component i, the $x_i f_i^{\Delta}$ is equals to one and therefore, the fugacity is equals to $H_{i,x}$.

There is a ratio:

$$\frac{\hat{f}_i}{f_i^{\bullet}} = \frac{\hat{f}_i}{H_{i,x}} = x_i f_i^{\nabla}$$

According to this ratio and the Eq. (2), the chemical potential of component *i* can be expressed as:

$$\mu = \mu_i^{\infty} + RT\ln(x_i f_i^{\nabla}) \qquad (4)$$

In the given equation provided above, μ_i^{∞} represents the chemical properties of component *i* in the standard state on the basis of the unsymmetrical convention and the mole fraction scale. On the molality scale, the activity of component *i* is $m_i\gamma_i$, in which γ_i represents the molal activity coefficient. The fugacity of component *i* in the solutions can be given by $\hat{f}_i = m_i\gamma_i H_{i,m}$, in which $H_{i,\,m}$ is the molal Henry's law constant for the component *i*.

The standard state is defined as the hypothetical ideal one molal solution. In the standard state, $m_i = m_{i,\,0} = 1$ mol (kg $H_2O)^{-1}$ and $\gamma_i = 1$. The standard state fugacity $f_i^{\bullet} = m_{i,0} H_{i,m}$. According to the Eq. (2), the chemical potential of the component *i* can be expressed as:

$$\mu = \mu_i^{\infty,m} + RT\ln\left(\frac{m_i}{m_{i,0}}\gamma_i\right) \qquad (5)$$

In the above equation, $\mu_i^{\infty,}$ 'epresents the chemical potential of component *i* in the standard stated on the basis of unsymmetrical convention and the molality scale. Usually, $m_{i,0} = 1$ mol (kg $H^2O)^{-1}$ is derived from the expression. The chemical potential of an ion does not depend on the definition of the standard state and the concentration unit. Therefore, it can be expressed in any of the three different ways like:

$$\begin{aligned} \mu &= \mu_i^0 + RT\ln(x_i, f_i) \\ &= \mu_i^{\infty} + RT\ln(x_i, f_i^{\nabla}) \\ &= \mu_i^{\infty,m} + RT\ln(m_i\gamma_i) \end{aligned} \qquad (6)$$

By re-arranging the first two terms on the right in Eq. (6), we can obtain,

$$\frac{\mu_i^0 - \mu_i^{\infty}}{RT} = \ln\frac{f_i^{\nabla}}{f_i} \qquad (7)$$

At the infinite dilution $f_i^{\nabla} = 1$, and so, the expression can be further written as:

$$\frac{\mu_i^0 - \mu_i^\infty}{RT} = -\ln f_i^\infty \quad (8)$$

In the equation given above, f_i^∞ represents the symmetrical, rational activity coefficient of ion i at infinite dilution. By combining the above two equations, we can obtain a correlation between the rational, symmetrical activity coefficient and the rational unsymmetrical activity coefficient.

$$f_i^\nabla = \frac{f_i}{f_i^\infty} \quad (9)$$

By further combining the terms of Eq. (6), we can get

$$\frac{\mu_u^0 - \mu^\infty}{RT} = \ln\left(\frac{v_i n_s \gamma_i (n_w + n_s \sum v_i)}{n_w M_w v_i n_s f_i}\right) = \ln\left(\frac{\gamma}{x_w f_i M_w}\right) \quad (10)$$

M_w represents the molar mass of water (in kg mol^{-1}). The reference molality 1 mol (kg H_2O)$^{-1}$ makes the expression dimensionless in the parenthesis. At infinite dilution, x_w, and γ_i both are close to unity, and f_i approaches f_i^∞. Then the expression can be obtained as:

$$\frac{\mu_i^0 - \mu_i^{\infty,m}}{RT} = \ln(M_w f_i^{\infty,m}) \quad (11)$$

By inserting this equation in the above Eq. (10) and combining with (9), we can deduce the following expression relating the molal activity coefficient to the rational symmetrical and the rational unsymmetrical activity coefficients:

$$\gamma_i = \frac{f_i}{f_i^\infty} x_w = f_i^\nabla x_w \quad (12)$$

From the Eq. (8) and 11, it follows that:

$$\frac{\mu_i^0 - \mu_i^{\infty,m}}{RT} = \ln M_w \approx 4.0165 \quad (13)$$

$\mu_i^\infty$and $\mu_i^{\infty,m}$ both represent the chemical potential of ion i on the basis of the unsymmetrical convention but refer to two different concentration scales. The ratio in between the mole fraction and the molality of ion i which is given by:

$$\frac{x_i}{m_i} = \frac{\dfrac{n_s v_i}{n_w + n_s \sum v_i}}{\dfrac{n_s v_i}{n_w M_w}} = \frac{n_w M_w}{n_w + n_s \sum vi} = x_w M_w \tag{14}$$

At infinite dilution, the ratio is equal to M_w, whereas f_i^∇ and γ_i both are equals to one. Therefore, the extrapolation of measured chemical potentials to infinite dilution yields the difference $RT\ln M_w$, which is already been mentioned in Eq. (13), and it depends on whether the mole fractions or molality are used as concentration unit.

4.7.2. Thermal Properties

In context to the temperature at constant pressure and composition, one obtains:

$$\left[\frac{\partial \ln f_w}{\partial T}\right]_{p,x} = \left[\frac{\partial\left(\dfrac{\mu_w}{RT}\right)}{\partial T}\right]_{P,x} - \left[\frac{\partial\left(\dfrac{\mu^0_w}{RT}\right)}{\partial T}\right]_P = -\frac{H_{w,x}}{RT^2} + \frac{H^0}{RT^2} = -\frac{L_{w,x}}{RT^2} \tag{15}$$

where, $H_{W,X}$ is the partial molal enthalpy of water at the particular composition; H_W^o is the molal enthalpy of water in the standard state; $L_{W,X}$ is the so-called relative molal enthalpy of water at the specified composition.

A similar kind of result is obtained for the ions by differentiation in respect to temperature:

$$\left[\frac{\partial \ln f_i^\nabla}{\partial T}\right]_{p,x} = \left[\frac{\partial\left(\dfrac{\mu_i}{RT}\right)}{\partial T}\right]_{P,x} - \left[\frac{\partial\left(\dfrac{\mu^\infty_i}{RT}\right)}{\partial T}\right]_P = -\frac{H_{i,x}}{RT^2} + \frac{H_i^\infty}{RT^2} = -\frac{L_{i,x}}{RT^2} \tag{16}$$

where, H_i^∞ is the partial molal enthalpy of ion 'I' in the standard state.

The same result would have been obtained by differentiation of the logarithm of the molal activity coefficients.

Equation (13), which has been mentioned above, shows that the difference between the two standard state chemical potentials divided by RT is independent of temperature. The partial molal enthalpy at infinite dilution is the same in these two standard states.

The relative enthalpy at constant composition of an electrolyte solution is given by:

$$L_x = \sum_{w+i} n_i L_{i,x} = n_s L_{\phi,x} \tag{17}$$

The term 'w+i' means that the summation is over water and ions.

$L_{\phi, x}$ is the apparent relative molal enthalpy of the salt at the particular composition. For systems where chemical relaxation because of the process of hydrolysis or association or dissociation reactions take place, the enthalpy of the electrolyte solution is much more useful.

$$H_x = \sum_{w+1} n_i H_{i,x} = n_s L_{\phi,x} + \sum_i n_i H_i^\infty + n_w H_w^0 \tag{18}$$

Mentioned above is the general equation (Eq. 18) from which the partial molal enthalpies of water and salt can be calculated by numerical differentiation:

$$H_w = \left[\frac{\partial H_x}{\partial n_w}\right]_{T,n=Eq} = L_w + H_w^0 \tag{19}$$

where, the subscript n is equal to 'equilibrium condition' which means that mol of salt is constant, though the composition is the equilibrium composition; Lw is the relative molal enthalpy of water.

The partial molal enthalpy (H_s) of salt and the relative molal enthalpy (L_s) of salt can be given by the equation:

$$H_s \aleph \left[\frac{\partial H_x}{\partial n_w}\right]_{T,n=Eq} \quad L_s \quad \sum_i v_i H_i^\infty \tag{20}$$

where, the v_i is known as stoichiometric coefficient. It shows that the summation comprises the ions of the original salt, not only the ions that are formed the chemical relaxation. The relative enthalpy 'L' of the solution and the relative molal enthalpy of the salt are then given by the equation below:

$$L = n_w L_w + n_s L_s = n_s L_\phi \tag{21}$$

The temperature derivative of *L* is the relative heat capacity *J* which can be expressed as

$$J = \left[\frac{\partial L}{\partial T}\right]_{n=Eq} = n_s \left(C_{p,\varphi} - \sum_i V_i C_{p,i}^\infty\right) \tag{22}$$

$C^\infty_{p,i}$ represents the standard state heat capacity of ion *i*. $C_{p,\varphi}$ represents the apparent molal heat capacity of the salt. It can be obtained from the apparent relative molal enthalpy of the salt by using the formula mentioned below:

$$C_{p,\varphi} = \left[\frac{\partial L\varphi}{\partial T}\right] + \sum_i V_i C_{p,i}^{\infty} \tag{23}$$

As per the observations, it can be said that the relative molal enthalpy can be defined by using Eq. (21) and the relation heat capacity can be defined by using the Eq. (22) and both are equal to zero in the case of the standard state of the ions only if no hydrolysis/association/dissociation reactions occur. The activity coefficient is equals to one in the standard state, but the equilibrium constants maintain their value. The heat capacity given as C_p of an electrolyte solution is the sum of standard state heat capacities of the components and the relation heat capacity which is given as:

$$C_p = \sum_{w+1} n_i C_{p,i}^0 + J = n_w C_{p,w}^0 + n_s C_{p,\varphi} \Rightarrow C_{p,\varphi} = \frac{C_p - n_w C_{p,w}^0}{n_s} \tag{24}$$

Which is able to explain the term ‘apparent’ and further establishes an important relation in between the apparent molal heat capacity of a salt and the heat capacity of a salt solution.

4.8. THERMODYNAMIC PROPERTIES OF ELECTROLYTES IN AQUEOUS SOLUTION OF GLYCINE

There are certain properties which are found in the aqueous solutions of amino acids, such as ultrasonic study and transport properties which are very useful in order to study different types of interactions in solutions.

However, the characterization of these interactions can assist while undertsanding the thermodynamic stability of proteins and their unfoldign behavior. There are electrolytes which tend to dissolve in water and they have been cclassified as structure makers or structure breakers, which depends on the charge density.

It has been identified that the ions having low charge density are the structure breakers and the ions having high charge density are the net structure makers and behave much opposite to the other one. It is an interesting field for exploring the ionic processes along with the solution of strong electrolytes.

Also, in aqueous solution of glycine, the interactions bof electrolytes and their temperature dependence of these interactions play an important role in understanding the nature of action of bioactive molecules or the thermodynamic behaviors of biochemical process considering the body system

In fact, there are extensive study related to the thermodynamical property of electrolytes in aquesous or non-aqueous solutions but there are only a few studies which are done in aqueous glycine solutions.

4.9. CONCLUSION

A chemical compound which gets dissociated into ions ans so, becomes able to transport electric charge can be referred to as an electrolyte. In other words, it can be said that, an electrolyte is an electric conductor. This chapter mainly focuses on electrolytes and thermodynamic properties. In the first half of the chapter, it describes the concept related to electrolyte including its pairing with body fluids and batteries and also, some examples of electrolytes.

This part contains one table that mentions the parameters of the dielectric constant. In the second half of the chapter, thermodynamics of electrolyte solutions have been mentioned in which it provides knowledge about the Gibbs energy functions for real electrolyte solutions and also, the thermal properties along with some important equations.

The conclusion that can be drawn from the chapter is that the chapter provides insights about the thermodynamic properties of electrolyte which is important in order to study different types of interaction going on in the solutions.

REFERENCES

1. Chem.libretexts.org. (n.d.). *Electrolytes—Chemistry LibreTexts*. [online] Available at: https://chem.libretexts.org/Bookshelves/Inorganic_Chemistry/Supplemental_Modules_(Inorganic_Chemistry)/Chemical_Reactions/Chemical_Reactions_1/Electrolytes (accessed on 11 March 2020).
2. Chemicool.com. (n.d.). *Definition of Electrolyte—Chemistry Dictionary*. [online] Available at: https://www.chemicool.com/definition/electrolyte.html (accessed on 11 March 2020).
3. Kaj, T., (2009). *Electrolyte Solutions: Thermodynamics, Crystallization, Separation methods*. [ebook] p. 3. Available at: https://backend.orbit.dtu.dk/ws/portalfiles/portal/131770579/ElectrolyteCompendium.pdf (accessed on 11 March 2020).
4. Kanhekar, S., Parwar, P., & K Bichile, G., (2009). *Thermodynamic Properties of Electrolytes in Aqueous Solution of Glycine*. [ebook] p. 1. Available at: http://nopr.niscair.res.in/bitstream/123456789/7236/1/IJPAP%2048%282%29%2095–99.pdf (accessed on 11 March 2020).
5. Majer, V., Sedlbauer, J., & Wood, R. H., (n.d.). *Chapter 4 Calculation of Standard Thermodynamic Properties of Aqueous Electrolytes and Non-Electrolytes*. [ebook] p. 2. Available at: https://www.kch.tul.cz/sedlbauer/chapter.pdf (accessed on 11 March 2020).
6. Shipley Rowlinson, J., Prausnitz, J. M., & Poling, B. E., (1999). *Liquid – Solutions and Solubilities*. [online] Encyclopedia Britannica. Available at: https://www.britannica.com/science/liquid-state-of-matter/Solutions-and-solubilities#ref51884 (accessed on 11 March 2020).
7. Thomsen, K., (1997). *Aqueous Electrolytes: Model Parameters and Process Simulation*. [ebook] p.3. Available at: http://www.phasediagram.dk/Thesis%20k%20thomsen%201997.pdf (accessed on 11 March 2020).
8. Zuber, A., Checoni, R., & Castier, M., (2015). Thermodynamic properties of nonaqueous single salt solutions using the q-electrolattice equation of state. Brazilian Journal of Chemical Engineering, [online] 32(3), 637–646. Available at: http://www.scielo.br/scielo.php?script=sci_arttext&pid=S0104–66322015000300003 (accessed on 11 March 2020).

CHAPTER 5

The Concept of Reversibility and Irreversibility in Thermodynamics

CONTENTS

The chapter of concept of reversibility and irreversibility in thermodynamic talks about basic significance of the reversibility and irreversibility. This chapter also mentions difference between irreversible and reversible process. This chapter also explains the thermodynamically reversible processes as sets of irreversible processes such as properties required, and the characterization. The chapter also explains, what are thermodynamically reversible processes must do such as the efficiency of heat engines, reversible engines are the most efficient, all reversible engines have the same efficiency, and absolute temperature. This chapter provides highlight on types of reversible process such as internally reversible process, and externally reversible process. This chapter is consisting of Irreversible process, the basic principle of Carnot engine, efficiency of the Carnot cycle, and some overall comments on entropy such as, basic significance of reversible and irreversible processes, and some example of reversible and irreversible processes.

5.1. INTRODUCTION

The thermodynamics of irreversible processes is a continuum theory of processes in matter with no reference to its molecular constitution. Nevertheless, the reductionist tradition of natural philosophy inevitably needs a molecular theoretical foundation.

This is needed in the form of a particulate theory of matter that is based on the concept of particles (for instance, atoms, and molecules) held presently in science.

The kinetic theory of fluids has been mainly developed in order to provide macroscopic phenomena and it also provides thermodynamics of reversible as well as irreversible processes in continuum matter with molecular theoretical foundations.

In such kind of theory, a suitable kinetic equation is useful for the distribution function of the system in the phase space or in the Hilbert space. On the other hand, in practice the solution of the kinetic equation is sought in such a manner so as to help understand at the thermodynamic level of description the macroscopic phenomena which is often experienced or observed in nature as well as in the laboratory.

In this way, the solution is a particular solution that is corresponding to the thermodynamic level of understanding of the natural phenomena of interest. Then, it is inevitable that in the first place, what is really meant by the thermodynamic level of description is examined.

During the course of the scientific training, thermodynamics of reversible processes has been taught. In addition to it, a great deal of the scientific reasoning and thinking processes is molded and it is influenced by equilibrium thermodynamics. This is particularly when the individual is concerned with macroscopic phenomena in continuum matter, but the subject is not very helpful when they face irreversible macroscopic processes.

In this way, the initial task in attempting to study the kinetic theory of matter is in developing a theoretical formalism for macroscopic phenomena in continuum matter on the foundations of the forms of laws of thermodynamics as proposed by their formulators presently.

Then, there is the development of a solution procedure for the kinetic equation for the distribution function of matter in the Hilbert space or the phase space. This depends on whether the quantum or classical mode of description is adopted for the development of the theory, in such a manner that the thermodynamics of irreversible processes is defined from the molecular perspective.

In this way, the individuals were motivated or encouraged in order to develop first a thermodynamic theory of irreversible processes. This theory of irreversible processes was developed without using the molecular picture of matter. In this effort, one should take the position that the first task is to develop a logical structure. This structure should be developed within the framework of the original form of the laws of thermodynamics as it is phrased by their formulators.

Since, it has been observed that the most difficult as well as unsettled part of the theory is concerned with the mathematical representation of the second law of thermodynamics. Close attention should be devoted to it, but it will be brief with the representation of the first law of thermodynamics and also the other conservation laws as well as the zeroth law regarding the temperature, since there is nothing new to add to them currently.

The applications of the theory that are developed for irreversible processes will be deferred later in the chapter. It will be assumed that the substance does not have an intrinsic angular momentum. Also, it is assumed that the system includes an r-component non-reactive mixture.

Thus, there is the exclusion of case of chemical reactions, but it is very easy to modify the theory to include the case of chemical reactions. After the completion of the formulation of the theory under the assumptions taken, it should be pointed out that where the required modifications must be made to involve the chemical reactions.

Since, it has been observed that the kinetic theory part of this work does not deal with fluids with an angular momentum, the irreversible thermodynamics of such fluids will not be considered.

5.2. DIFFERENCE BETWEEN IRREVERSIBLE AND REVERSIBLE PROCESS

Reversible process is defined as the sequence of modification or alterations that are carried out on the system through the course of its change from the initial to final state may be possibly reversed in an exact manner in a reversible process.

This is only possible when the series of changes are carried out very slowly that is in many smaller steps on the system during its transformations from initial to final state. By doing in such a way, each of its intermediate state will be in equilibrium or balance with its surroundings. Under such kind of conditions, the initial as well as the final states of the system become reversible completely.

For instance, a certain amount of heat is absorbed when ice melts. The water that is formed can be transferred back to ice. This is possible if the same amount of heat is removed from it. This shows that many of the reversible processes are non-spontaneous processes also.

Irreversible process is defined as the process which cannot be retraced to the initial state with no permanent change in the surroundings. It has been observed that many of the spontaneous processes are irreversible in nature.

For instance, Biological ageing is an irreversible process. Water that is flowing down a mountain on its own accord is an irreversible process.

Some of the attributes of thermodynamically reversible as well as irreversible processes are compared as Table 5.1.

Table 5.1. Difference between Reversible Process and Irreversible Process

Reversible process	Irreversible process
It is a slow process going through a series of smaller stages with each stage maintaining equilibrium between the system and surroundings.	In this process the system attains final state from the initial state with a measurable speed. During the transformation, there is no equilibrium maintained between the system and surroundings.

A reversible process can be made to proceed in forward or backward direction.	Irreversible process can take place in one direction only.
The driving force for the reversible process is small since the process proceeds in smaller steps.	There is a definite driving force required for the progress of the irreversible process.
Work done in a reversible process is greater than the corresponding work done in irreversible process.	Work done in an irreversible process is always lower than the same kind of work done in a reversible process.
A reversible process can be brought back to the initial state without making a change in the adjacent surroundings.	An irreversible process cannot be brought back to its initial state without making a change in the surroundings.

5.3. THERMODYNAMICALLY REVERSIBLE PROCESSES AS SETS OF IRREVERSIBLE PROCESSES

5.3.1. Properties Required

The following are the characteristics which are generally linked with a reversible process in the field of thermodynamics. They will create the basis of a characterization free of manifest paradox. They are the processes that can continue in either the forward direction as well as in the reverse direction, the time, when their thermodynamic forces are in almost in perfect balance.

If heat Qf is acquired with the help of the system and work W_f is done with the help of the system in the forward process, then heat can be written as

$$Q_r = -Q_f$$

Which is acquired by the system and work $W_r = -W_f$ is done with the help of the system in the reverse process.

If a space of equilibrium states is available, then the process can be represented with the help of a continuous curve in the space.

The rate at which work done took place, can be written as

$$dW = \mathbf{\Sigma i}\ Xi\ dxi \qquad (1)$$

This expression (i) is derived with the help of a standard statement of the first law of thermodynamics

$dU = dQ - dW = dQ - \mathbf{\Sigma i}\, Xi\, dxi$ (2)

In which, U is the internal energy of the system and the paired variables X1 and x1, X2 and x2, etc. are pairs of generalized force (Xi) and displacement (xi) variables. Common pairings are pressure which is denoted by P and volume by V, surface tension as γ and surface area by A, along with the magnetic quantities as well as electric quantities like magnetic field and magnetic moment; and the strength of electric field and dipole moment. (Pippard, 1966. pp. 23–28.)

The operator D is commonly showed as marking a small or even infinitesimal variation in the quantity on which it acts. A mathematically cleaner reading is to locate a path parameter λ with respect to the process and identify the operator as $d = d/d\lambda$.

Here, the term "driving force" depicts the variations in the temperature and these kinds of generalized forces. These variations or differences in the temperature is defined in the non-equilibrium systems. According to the, Fourier's equation for the dynamics of heat transport depicts the temperature gradients as a force that is main reason behind heat flow.

According to the Navier Stokes equation, which employs the pressure gradients as a force that determines the momentum in the fluids. More commonly, external to thermodynamics, pressure as well as surface tension are well-defined in theories of the mechanics of fluids; and electric and magnetic field are well-defined in electrodynamics.

5.3.2. The Characterization

These properties are all captured by the following:

Definition: The label "thermodynamically reversible process" depicts a set of irreversible processes in a thermal system, which is delimited with the help of the set of equilibrium states in (d) such that:

(a) Each and every process can exchange heat or work with its surroundings, due to the imbalanced driving forces (which are variations in the temperature, generalized forces).

(b) The processes can be separated into a "forward" and a "reverse" set, in such a way that, the total heat acquired, and the total work done have opposite signs in the two sets.

(c) In each set, there are processes in which the net driving forces are subjectively small. With respect to the, generalized forces, the net driving force is the fluctuation the generalized force and the force in the surrounding system that counters it.

(d) Under the limit of these net driving forces going to zero, the states of both forward processes as well as reverse processes methodologies the same set of equilibrium states and these types of states form a curve in equilibrium state space.

(e) The limiting values of heat acquired, and work done with the help of the forward process are Qf and W_f; and with the help of the reverse process Q_r and W_r; and they satisfy

$Q_f = -Q_r$ and $W_f = -W_r$

(f) These limiting quantities of heat and work can be calculated at any phase of the process, with respect to these calculated values, with the help of the integration of the relations (i) and (ii) along the curve of the equilibrium states in equilibrium state space.

On the other hand, they may come subjectively close, none of the states of the irreversible processes are exactly equilibrium states. And, if not, the processes cannot complete in any finite time. The set is represented in Figure 5.1.

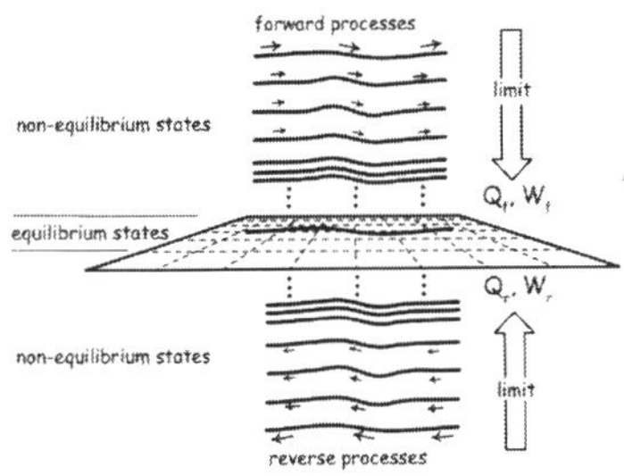

Figure 5.1. *Set of processes forming a reversible process.*

The definition needs an additional statement or postulation if it is to be applied in the field of thermodynamics:

There is a thermodynamically reversible process for any curve in equilibrium state space. Whether the present assumption is true which relies on the specific dynamical theory that governs the alteration or modifications in the thermal systems. Nevertheless, it can fail.

At the end, this evaluation can be utilized with respect to the heat engines, from the time, they are consisting of a series of thermal processes. A reversible heat engine is developed with the help of the corresponding series of the thermodynamically reversible processes.

5.4. WHAT THERMODYNAMICALLY REVERSIBLE PROCESSES MUST DO

It can be seen that the categorization of the thermodynamically processes which are suitable and also, they provide the assistance in rederivation of the results that are derived conventionally with the help of the notion thermodynamically reversible processes. According to the Sommerfeld's remarks Sommerfeld's (1956, p. 19) that suggests the aspects of the rederivations when he wrote:

Reversible processes are the most necessary aspects in the field of thermodynamics, and the reason behind that is, there are definite equations that can be acquired only with the help taking into the consideration of the reversible changes. On the other hand, irreversible changes or modifications can only be explained with the help of the inequalities when the equilibrium thermodynamics is applied.

With respect to the remarks of Sommerfeld, what he does not mentioned is that, if an individual has sufficient amount of inequalities, the derivations are very important, but there are some complexions with respect to the application of the inequalities.

5.4.1. The Efficiency of Heat Engines

According to the Carnot's (1824) essay, which is an extraordinary innovation, was a means of making conclusions about some general constraints with respect to the effectiveness of the heat engines. Carnot pictured about the heat engines as deriving the mechanical effect with the help of passing the heat from a place which has higher temperature to the place having comparatively low temperature.

Carnot was able to depict that the greatest mechanical effect for any two these types of places was generated with the help of the heat engine that could be put into use with respect to the reverse and that all such reversible engines which can performed equally well, as well as independently of their construction and whether their working material was air, steam or anything else.The original analysis of the Carnot was direct within the caloric theory of heat under the postulation that heat is conserved through the course of its passage from hot to cold. There is another subsequent analysis that is done by Clausius (1851) and Thomson (1852) altered the analysis done by the Carnot with the help of requiring that the work produced with the help of a heat engine, which as an outcome, derived from the conversion of some of the heat into the work.

The results of the Carnot are stated with respect to the internally contradictory notion of a reversible heat engine that could transfer heat without the variation in the temperature. A heat engine is a series of a thermodynamically reversible processes that draws and releases the heat from and to a hot place and cold place. On the other hand, through the course of the conversion of the heat energy into the work. Integrating them, this engine is just a single lengthy thermodynamically reversible process.

The efficiency η of a heat engine is defined in the common way $\eta = W/Q$, where W is the work delivered with the help of the engine and Q the heat drawn from the hot place. A reversible heat engine is consisting of a set of irreversible processes that produces a reversible process.

Its efficiency η (rev) is defined as the ratio of the limiting values W (f) and Q (f) for W and Q. There is no postulation and supposition that any individual engine or processes in the set can realize this efficiency.

5.4.2. Reversible Engines Are the Most Efficient

There is no irreversible heat engine working between the same temperature can have a greater effectiveness as compared to the effectiveness of a reversible engine. In order to show this, assume for the reason of reduction that have an irreversible engine which have the efficiency η (irr) = W (irr)/Q (irr), which is greater as compared to the reversible engine.

η (rev) = W (f)/Q (f) = W (r)/Q (r)

That is

η (irr) > η(rev)

Noticeably, do not assume that any real heat engine can function with the efficiency **η (rev)**. Nevertheless, the definition of a reversible process permits that an irreversible heat engine can come subjectively close in its operation.

In addition to this, since the process is reversible, these irreversible heat engines are found in both the "forward" as well as "reverse" sets of irreversible processes. That is, there is a series of processes in the reverse set that produces heat Q to the hot place and utilizes work W, such that

b

Where the non-zero ε can be brought arbitrarily close to zero with the help of a suitable selection of processes in the set. The amplitude of the functions of the two engines can be measured in such a way that W (irr) =

W and combination of the two engines in the general way: work from the irreversible engine is utilized in order to operate the reversed engine, as shown in Figure 5.2.

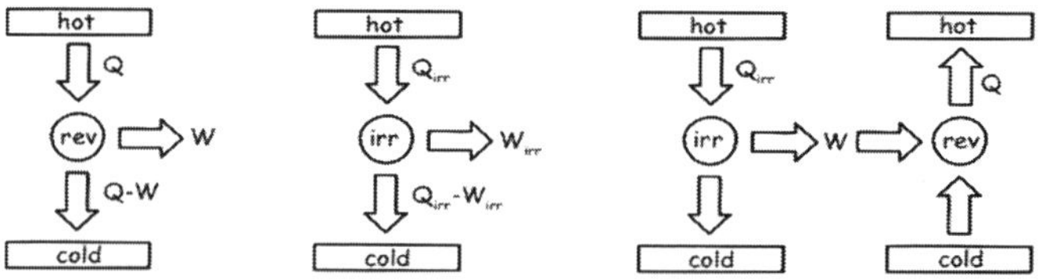

Figure 5.2. *Coupled reversible and irreversible heat engines.*

That is, the equation can be written as $W = Q (\eta_{rev} - \varepsilon) = Q_{irr}\, \eta_{irr} = W_{irr}$, so that

$$\mathbf{Q/Q_{irr} = \eta_{irr}/\eta_{rev} - \varepsilon > 1}$$

The second inequality arose at the time, we can make ε subjectively small and it has been assumed that η (irr) > η(rev). the net impact of the integrated operations of the two engines is easy to take heat Q – Q (irr) from the cold place and carry it to the hot place. If Q/Q (irr) > 1, in this way, this quantity of heat Q-Q (irr) is a positive quantity.

This is a disruption of the Clausius form of the second law of the thermodynamics, for which there is a process whose single effect is to transfer or carry the heat from the place having low temperature to the place of high temperature without the net utilization of the work. This denial completes the reduction process. No irreversible heat engine can be more effective as compares to the effectiveness assigned to a reversible engine.

5.4.3. All Reversible Engines Have the Same Efficiency

Identical implications permit to show that all the reversible heat engines which are operating between the same temperatures have the same effectiveness or efficiency. As before, a reversible heat engine is a set of processes. There are no postulations that any kind of process realizes the efficiency η (rev).

Assumptions for the reasons of reduction that there have two these kinds of reversible engines, A, and B, that functions or operates with various efficiencies, with the first more effective as compared of the second on. That is,

$$\eta A = W (A, f)/Q (A, f) > W (B, f)/Q (B, f) = \eta (B)$$

In this way, people can find out the real engines that functions at efficiencies within arbitrarily small, non-zero ε_A and ε_B of these unrealizable efficiencies, with quantities of work and heat:

$W_A/Q_A = \eta_A - \varepsilon_A$ $\qquad\qquad$ $W_B/Q_B = \eta_B - \varepsilon_B$

In order to measure the operations of the two engines, in this way, W = WA = WB. They can now be combined, in such a way that work from the engine A is used in order to drive the second less efficient engine B in reverse, as shown in the Figure 5.3.

Engine A draws heat Q_A from the place of high temperature and engine B returns heat Q_B to it. There is a net transport of heat Q_B-Q_A from the place of low temperature to the place with high temperature, with no network exchanged with the environment.

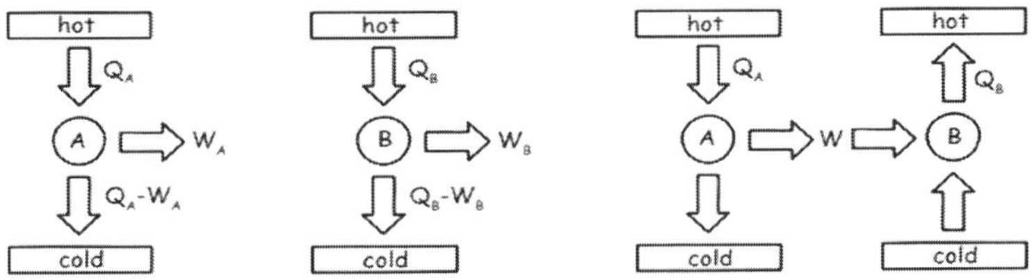

Figure 5.3. *Coupled reversible heat engines.*

The equalities $W = W_A = W_B$ permits to recover

$W = W_A = (\eta_A - \varepsilon_A)\, Q_A = (\eta_B - \varepsilon_B)\, Q_B = W_B$

In this way, the equation can be written as:

$Q (B)/Q (A) = (\eta_A - \varepsilon_A)/(\eta_B - \varepsilon_B) > 1$

The second inequality arose since we make ε_A and ε_B as small as we like, on the other hand, η_A and η_B are fixed in the value with $\eta_A > \eta_B$. It follows that

The second inequality arises since we can make ε_A and ε_B as small as we like, while η_A and η_B are fixed in value with $\eta_A > \eta_B$. It follows that $Q_B - Q_A > 0$. The net effect of the cycle is to transfer heat from hot too cold without any net consumption of work. The Clausius form of the second law of thermodynamics is violated and the reduction is complete.

Analogous arguments give the same consequences for refrigerators or the heat pumps. They are heat engines that functions in the reverse direction, with the help of work in order to move heat from a place with low temperature to a place with high temperature.

As an alternative of efficiency, we have the coefficient of performance (COP), which is the ratio of work consumed to heat drawn from the cold place. No irreversible heat pump outperforms a reversible heat pump; and all heat pumps operating between the equal temperatures have the same COP.

5.4.4. Absolute Temperature

The consequences of the last section of this topic, takes into the direction towards the existence of the absolute temperature scale. In order to recover it, it has been considering two reversible heat engines integrated together as shown in figure 5.4, on the other hand, before the quantities of heat and work which is mentioned are constrained values, unrealized with the help of any kind of heat engine.

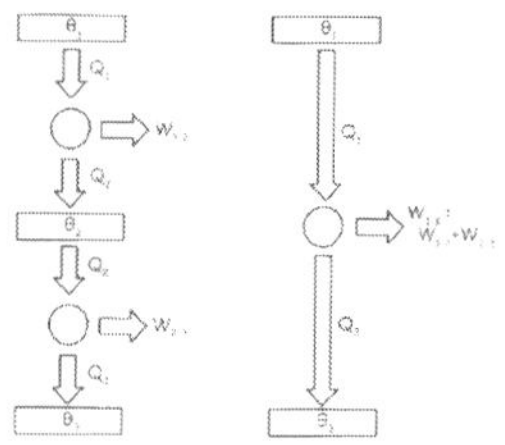

Figure 5.4. *Composition of heat engines in the derivation of absolute temperature.*

The first draws heat Q_1 from the highest temperature θ1 and releases heat Q_2 in order to intermediate temperature θ_2, on the other hand, creating work W_{1-2}. Temperature is written here as θ in order to depict that a definite scale for the temperature is not yet assumed.

The efficiency $\eta = W_{1-2}/Q_1$ of this engine is a function of the temperature θ_1 and θ_2 only. It follows that the ratio of heat drawn to discharged heat is $Q_2/ Q_1 = (Q_1-W_{1-2})/Q^1$ is also a function of θ_1 and θ_2 only. In this way, the equation can be written as

$$F(\theta 1, \theta_2) = Q_2/Q_1 \qquad (3)$$

The second heat engine draws heat Q2 from the intermediate temperature θ2 and release the heat Q_3 to the lowest temperature θ3. We have analogously for it

$$F(\theta_2, \theta_3) = Q_3/Q_2$$

When the two cycles are integrated, as shown in Figure mentioned above, form a single larger heat engine operating between the temperatures θ_1 and θ_3 with heats Q_1 and Q_3. It can be written as:

$F(\theta_1, \theta_3) = Q_3/Q_1$

Integrating these last three equations, the final equation can be written as

$F(\theta_1, \theta_3) = Q_3/Q_1 = (Q_3/Q_2).(Q_2/Q_1) = f(\theta_1, \theta_2).F(\theta_2, \theta_3)$

It now follows that, then equation will be

$$F(\theta_1, \theta_2) = f(\theta_1, \theta_3)/f(\theta_2, \theta_3) = (1/T_1)/(1/T_2) \quad (4)$$

where we introduce the absolute temperature T by definition as

$T(\boldsymbol{\theta}) = 1/F(\theta, \theta_3)$

In the end, integrating the equations (3) and (4), we have $Q_2/Q_1 = T_2/T_1$ so that

$$Q_1/T_1 = Q_2/T_2 \quad (5)$$

5.5. REVERSIBLE PROCESS

A thermodynamic process is said to be reversible if the process can return back in such a manner that both the system as well as the surroundings return back to their original states, without any other change anywhere else in the universe. This means that both system as well as surroundings are returned to their initial states at the end of the reverse process.

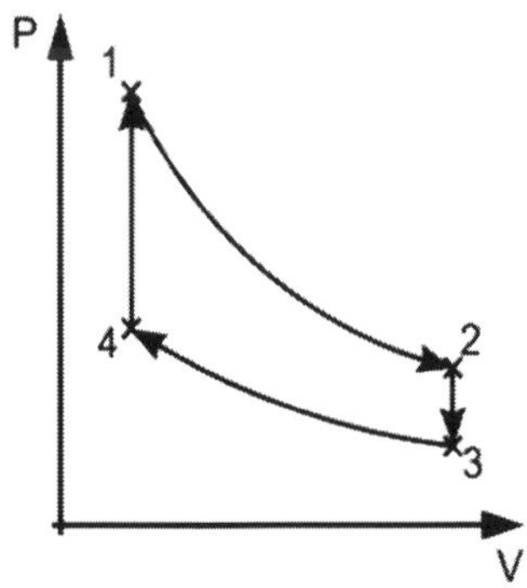

Figure 5.5. *An illustration of reversible process.*

In Figure 5.5, it is shown that the system has undergone a change that is from state 1 to state 2. It has been observed that the reversible process can reverse completely and there is no trace left which can show that the system had gone through thermodynamic change. All the changes in state that takes place in the system are in thermodynamic equilibrium with each other during the reversible process.

5.5.1. Internally Reversible Process

The process is said to be internally reversible if no irreversibility's takes place within the boundaries of the system. In these kinds of processes, a system goes through a series of equilibrium states. The system passes through exactly the same equilibrium states while returning to its initial state when the process reverse.

5.5.2. Externally Reversible Process

In externally reversible process, there are no irreversibility's that takes place outside the system boundaries during the process. For example, the heat transfer between a reservoir as well as a system is an externally reversible process if the surface of contact that is between the system and reservoir is at the same temperature.

A process can be reversible only when its satisfying two conditions. These conditions are:

- Dissipative force must be absent; and
- The process should take place in infinite small time.

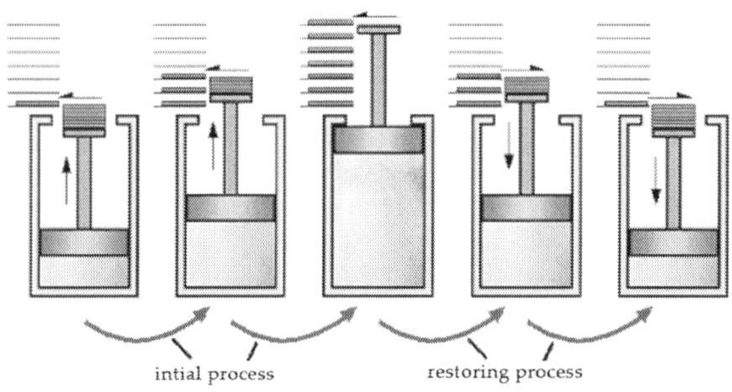

Figure 5.6. *Reversible process and irreversible process.*

5.6. IRREVERSIBLE PROCESS

The irreversible process is also known as the natural process. This is because all the processes that are taking place in nature are irreversible processes. The natural process takes place because of the finite gradient between the two states of the system.

For example, heat that flow between two bodies takes place because of the temperature gradient that is between the two bodies; in fact, this is the natural flow of heat. Likewise, water flows from high level to the low level and current moves from high potential to low potential, etc.

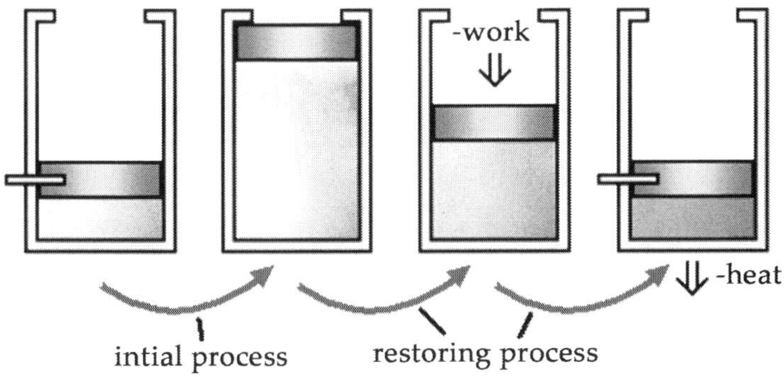

Figure 5.7. *An illustration of irreversible process.*

Irreversible processes are a consequence of straying away from the curve, thus reducing the amount of overall work done. An irreversible process is defined as a thermodynamic process that departs from equilibrium. In the context of pressure and volume, it takes place when the pressure (or the volume) of a system alters dramatically and instantly that the volume (or the pressure) do not have the time to reach equilibrium.

One of the classic examples of an irreversible process is permitting a certain volume of gas to release into a vacuum. By emitting pressure on a sample and permitting it to occupy a large space, the system as well as the surroundings are not in equilibrium during the process of expansion.

Though, there is a need of a significant work, with an equivalent amount of energy dissipation as heat flows to the environment. This is in order to reverse the process.

There are some key points about the irreversible process:

1. The initial state of the system and surroundings cannot be restored from the final state in the irreversible process;
2. The various states of the system on the path of change from initial state to final state are not in equilibrium with each other During the irreversible process;
3. The entropy of the system increases decisively, and it cannot be decreases back to its initial value during the irreversible process; and

4. The phenomenon of a system that is undergoing irreversible process is known as irreversibility.

5.7. THE CARNOT ENGINE

The Carnot engine is a hypothetical heat engine. This engine is based on a completely reversible cyclic process. The cycle includes four separate steps as shown in the pV diagram (Figure 5.8):

- Reversible isothermal expansion at TH;
- Reversible adiabatic expansion (TH→TL);
- Reversible isothermal compression at TL; and
- Reversible adiabatic compression (TL→TH).

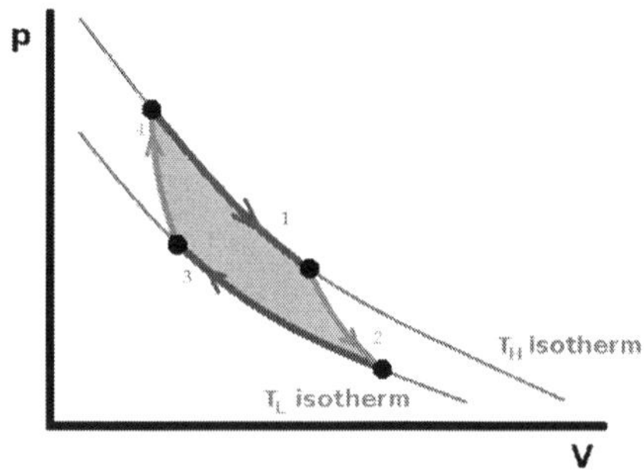

Figure 5.8. *An illustration of Carnot cycle.*

A cycle between two isotherms is operated by the engine at different temperature. The two isothermal steps are associated by adiabatic ones. Here (by definition) no heat is exchanged. Thus, the temperature must change (in the same manner as it does in the adiabatic throttle experiment).

Carnot's theorem stated that, for a given temperature difference,

- No irreversible engine can be more efficient; and
- All reversible engines have the same efficiency.

In order to validate this, consider two Carnot engines which is integrated back to back, that is, the work done with the help of the engine on the left is fed into the one on the right, which plays role of a pump. If the Carnot heat pump were not effective as compared to the heat engine, it would pump more heat from the cold reservoir to the hot one as compared to the flows through the heat engine in the opposite direction.

This process would disobey the statement given by the Clausius in the second law. Since all of the processes are irreversible in nature, the action of the both engines could be reversed, and the same argument would apply the other way. In this way, both of the engines have the identical efficiency if they are both engines are reversible.

Among two engines, if only one of the engines is reversible, it needs to be well thought out as an individual, what would occur if one of the engines was more effective.

If the irreversible heat engine was more efficient, it would take less amount of heat from TH and release less amount of heat at TL, again, as an outcome, with respect to the Clausius violation since the reversible heat pump would transfer more amount of heat up as compared to the amount of heat that flows down with the help of the heat engine.

Whereas, if the irreversible heat pump ware more effective, in this way, the amount of work done, W (cyc), received from the reversible heat engine would permit it to carry more amount of heat from the reservoir at TL as compared to the amount of heat fed into it from the heat engine.

In this situation, again, violating the Clausius statement. In this way, either way, the reversible component must be more effective one.

According to the Carnot's theorem, which is just another form of the second law of thermodynamics.

5.7.1. Efficiency of the Carnot cycle

Given that all of the reversible processes operating between the same two temperature are equally effective, it is easy to make arbitrary assumptions about the details of the system in order to evaluate the reversible efficiency.

It is easy to evaluate the effectiveness if the system that is consisting an ideal gas since we know the relationship between the state variables in that case.

Since the first step of the Carnot cycle is isothermal, the heat, QH is transferred from the hot reservoir equals the work done with the help of system through the course of the isothermal expansion. With the help of substituting the pressure according to the ideal gas law and integrating over the change volume, then the equation can be written as

$$\begin{aligned} Q_H &= W_{ab} \\ &= \int_a^b p\,dV \\ &= \int_{V_a}^{V_b} \frac{nRT_H}{V}\,dV \\ &= nRT_H \ln\frac{V_b}{V_a} \end{aligned}$$

and, in the same vein, the heat transferred from the system to the cold reservoir is

$$Q_L = W_{cd} = nRT_L \ln\frac{V_d}{V_c}$$

Note that the QL has a negative value (in line with the sign conventions introduced earlier) since Vd < Vc. To evaluate the efficiency of a reversible engine, we can start from the definition of efficiency:

$$\eta_{rev} = 1 - \frac{|Q_L|}{Q_H} = 1 - \frac{\left|T_L \ln\frac{V_d}{V_c}\right|}{T_H \ln\frac{V_b}{V_a}} = 1 - \frac{T_L \ln\frac{V_c}{V_d}}{T_H \ln\frac{V_b}{V_a}}$$

5.8. IRREVERSIBILITY, ENTROPY CHANGES, AND "LOST WORK"

Considering a system in contact with a heat reservoir through the course of a reversible process. If there is heat Q absorbed with the help of a reservoir at a temperature T, the shift in the entropy of the reservoir is Delta S = Q/T. Commonly, the reversible processes are assisted with the help of heat exchange that take place at the various temperatures.

In order to analyze these kinds of data, we can visualize a series of the heat reservoirs at various temperature, in such a way that, through the course any infinitesimal portion of the cycle there will not be any heat transferred over a finite variation in the temperature.

Through the course of the infinitesimal portion, the amount of heat dQ (rev) will be carried out between the system and one of the reservoirs is at T. If dQ (rev) is absorbed with the help of the system, the change in entropy of the system, can be written as

dS (system) = dQ (rev)/T

The change in entropy of the reservoir can be written as

dS (reservoir) = dQ (rev)/T

The change in total entropy of system plus surroundings can be written as

dS (total) = dS (system) + dS (reservoir) = 0.

This is also true if there is a quantity of heat rejected by the system.

At the end, with respect to a reversible process, no alteration take place in the total entropy produced, that is, the entropy of the system plus the entropy of the surroundings is

Delta S (total) = 0.

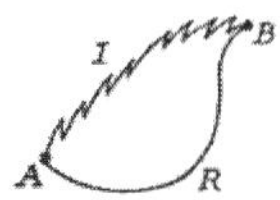

Figure 5.9. *Irreversible and reversible state changes.*

Now, for an irreversible process, carrying out the similar kind of analysis, which takes the system between the same specified states as in the reversible process. This is shown in Figure 5.9, in which I and R depicts the irreversible processes. With respect to the reversible process, the system receives heat dQ and does the work, which is dW. The change in the internal energy (U) with respect to the irreversible process can be written as

dU = dQ – dW (Always true – first law)

with respect to the reversible process, the equation can be written as

dU = TdS – dW (rev)

due to the state change is the same in the two processes, the alteration in internal energy is the same. Equating the changes in internal energy in the above two expressions yields, then the equation will be

dQ (actual) – dW (actual) = TdS – dW (rev)

The subscript "actual" refers to the actual process (which is irreversible in nature). The change in entropy is linked with the state change, can be written as

$$dS = \frac{dQ_{actual}}{T} + \frac{1}{T}\left[dW_{rev} - dW_{actual}\right] \qquad (6)$$

If the process is not reversible, we obtain less work as compared to a reversible process, where dW (actual) < dW (rev), in this way, with respect to the irreversible process, the equation will be dS > dQ (actual)/T

There is no equality between the entropy change dS and the quantity dQ/T for an irreversible process. The equality is only applicable for a reversible process.

The change in entropy for any process that leads to a transformation between an initial state "a" and a final state "b" is therefore

$$\Delta S = S_b = S_a \geq \int_a^b \frac{dQ_{actual}}{T},$$

where dQ (actual) is the heat exchanged in the actual process. The equality only applies to a reversible process.

The difference dW (rev) – dW (actual) represents work we could have obtained but did not. It is referred to as lost work and denoted by W (lost). In terms of this quantity we can write,

$$dS = \frac{dQ_{actual}}{T} + \frac{dW_{lost}}{T} \tag{7}$$

The content of Eq. (7) is that the entropy of a system can be changed or modified through two ways: (i) with the help of heat exchange, and (ii) with the help of irreversibilities. The lost work (dW (lost) in Eq. (7)) is always greater than zero, in this way, only way to reduce the entropy of a system is with the help of heat transfer.

In order to apply the second law, total change in entropy is considered (which is system plus surroundings). If the surroundings are a reservoir at temperature T, with which the system exchanges heat,

dS (reservoir) =dS (surrounding) = – dQ (actual)/T

The total entropy change is

$$dS_{Total} = dS_{system} + dS_{surroundings} = \left(\frac{dQ_{actual}}{T} + \frac{dW_{lost}}{T}\right) - \frac{dQ_{actual}}{T}$$

$$dS_{Total} = \frac{dW_{lost}}{T} \geq 0$$

The quantity (dW (lost)/T) is the entropy generated because of the irreversibility. Yet another way to state the distinction we are making is

$$dS_{system} = dS_{from\ heat\ transfer} + dS_{generated\ due\ to\ irreversible\ processes} = dS_{heat\ transfer} + dS_{Gen} \tag{8}$$

The lost work is also called dissipation and noted $d\Phi$. Using this notation, the infinitesimal entropy change of the system becomes:

dS (system) = dS (heat transfer) + dΦ / T

or

TdS (system) = dQ + dΦ

Equation (3) can also be written as a rate equation,

$$\frac{dS}{dt} = \dot{S} = \dot{S}_{\text{heat transfer}} + \dot{S}_{\text{Gen}} \qquad (9)$$

Either of Equation (8) or (9) can be understood to understand that the entropy of the system, which is denoted by S, is impacted by two factors: the flow of heat Q and the appearance of additional entropy, which is denoted by dS (gen), because of irreversibility. This additional entropy is zero when the process is reversible and always positive when the process is irreversible.

In this way, an individual can state that the system develops sources which develops the entropy through the course of an irreversible process. The second law asserts that sinks of entropy are impossible in nature, which is a more graphic way of saying that dS_{Gen} and $\dot{S}_{\text{Gen}}$ are positive definite (always greater than zero), or zero in the special case of reversible processes.

The term

$$\dot{S}_{\text{heat transfer}} \left(= \frac{1}{T}\frac{dQ}{dT}, \text{or} \frac{\dot{Q}}{T} \right)$$

Which is linked with heat transfer to the system, can be understood as a flux of entropy. The boundary is crossed by heat and the proportion of this heat flux to temperature can be described as a flux of entropy. There are no limitations on the sign of this quantity, and it can be said that this flux either contributes towards, or this flux drains away, the entropy of the system.

It has been observed that only this flux can affect the entropy of the system during a reversible process. It is suggested by this terminology that the entropy is interpreted as a kind of weightless fluid. The quantity of this fluid is conserved (like that of matter) during a reversible process. On the other hand, this fluid is not conserved during an irreversible process; it cannot disappear, but rather it is created by sources throughout the system. Though this interpretation should not be taken too factually or literally, it offers an easy mode of expression and is in the same category of concepts like those linked with the phrases "flux of energy" or "sources of heat." For instance, in fluid mechanics, this graphic language is very effective. Also, there should be no objections to copying it in thermodynamics.

5.8.1. EXAMPLES OF REVERSIBLE AND IRREVERSIBLE PROCESSES

Processes that are commonly idealized as reversible is consisting of:

- Frictionless movement;
- Restrained compression or expansion;
- Energy transfer as heat because of the infinitesimal temperature nonuniformity;
- Electric current flow with the help of a zero resistance;
- Restrained chemical reaction; and
- Mixing of two samples of the same substance at the same state.

Processes that are irreversible include:

- Movement with friction;
- Unrestrained expansion;
- Energy transfer as heat because of the large temperature non-uniformities;
- Electric current flow with the help of a non-zero resistance;
- Spontaneous chemical reaction; and
- Mixing of matter of different composition or state.

5.9. CONCLUSION

In the conclusion of thermodynamics, a reversible process is a type of process in which the direction can be "reversed" with the help of inducing infinitesimal alterations or modifications to some property of the system with the help of its surroundings. Throughout the entire reversible process, the system is in thermodynamic equilibrium with its surroundings.As an example of reversible process, a Carnot cycle process is not practically possible to build a thermodynamically reversible system, and the reason behind it is there is no such system which works with hundred per cent efficiency, that is work done by the system is hundred per cent. In simpler words, the it is impossible to create a system in which the conversion of heat into any kind of work with full efficacy.

REFERENCES

1. Bhaskar, R., & Pradeep, A. M., (n.d.). *Reversible and Irreversible Processes Concept of Entropy*. Bombay: Indian Institute of Technology, Bombay.
2. Encyclopedia Britannica, (2020). *Reversibility | Thermodynamics*. [online] Available at: https://www.britannica.com/science/reversibility (accessed on 11 March 2020).
3. Engineers edge, (2020). *Irreversible Process Thermodynamic | Engineers Edge | www.engineersedge.com*. [online] Engineersedge.com. Available at: https://www.engineersedge.com/thermodynamics/irreversible_process.htm (accessed on 11 March 2020).
4. Eu, B., (1998). Thermodynamics of irreversible processes. *Nonequilibrium Statistical Mechanics*, [online] pp.12–54. Available at: https://link.springer.com/chapter/10.1007/978–94–017–2438–8_2 (accessed on 11 March 2020).
5. Haresh, K., (2019). *What are Reversible and Irreversible Processes in Thermodynamics?*. [online] Bright Hub Engineering. Available at: https://www.brighthubengineering.com/thermodynamics/4616-what-are-reversible-and-irreversible-processes/ (accessed on 11 March 2020).
6. Nuclear Power, (2020). *Irreversible Process – Nuclear Power*. [online] Nuclear Power. Available at: https://www.nuclear-power.net/nuclear-engineering/thermodynamics/thermodynamic-processes/irreversible-process/ (accessed on 11 March 2020).
7. Physics LibreTexts, (2019). *4.2: Reversible and Irreversible Processes*. [online] Physics LibreTexts. Available at: https://phys.libretexts.org/Bookshelves/University_Physics/Book%3A_University_Physics_(OpenStax)/Map%3A_University_Physics_II_-_Thermodynamics%2C_Electricity%2C_and_Magnetism_(OpenStax)/04%3A_The_Second_Law_of_Thermodynamics/4.02%3A_Reversible_and_Irreversible_Processes (accessed on 11 March 2020).
8. Fowler, R. and Ackland, G., 2018. Lecture 2: Reversibility And Work. [ebook] Available at: <https://www2.ph.ed.ac.uk/~gja/thermo/lectures/lecture2.pdf> [Accessed 11 March 2020].
9. Rudi, W. W., (2005). *Entropy and Reversibility: Thermodynamics: Rudi Winter's Web Space*. [online] Users.aber.ac.uk. Available at:

http://users.aber.ac.uk/ruw/teach/215/entropy.php (accessed on 11 March 2020).

10. Twt.mpei.ac.ru. (n.d.). *Change of Entropy in Irreversible Processes.* [online] Available at: http://twt.mpei.ac.ru/TTHB/2/KiSyShe/eng/Chapter3/3–7-Change-of-entropy-in-irreversible-processes.html (accessed on 11 March 2020).
11. Web.mit.edu. (n.d.). Features of Reversible Processes. [online] Available at: https://web.mit.edu/16.unified/www/FALL/thermodynamics/notes/node34.html (accessed on 11 March 2020).
12. William, C. R., & Henry, C. P., (2020). 6.8 Some Overall Comments on Entropy. [online] Web.mit.edu. Available at: https://web.mit.edu/16.unified/www/FALL/thermodynamics/notes/node51.html (accessed on 11 March 2020).

CHAPTER 6

Stability of Thermodynamic System

CONTENTS

This chapter is about how the stability factor help in the thermodynamic system. Where stability is the mathematical condition in which if a slight disturbance is created in a system, but that disturbance does not impact on that system.

Then that type of system is said to be stable. In terms of thermodynamics, it is the heat energy of the system which can neither be consumed nor it be released. That system is said to achieve the stability. In this chapter there are several factors that are explained which are helping the system to achieve the stability. There are also some of the conditions that are affecting the stability of the thermodynamic system.

6.1. INTRODUCTION

There are many processes in the industry which practices with the unstable behavior. For example, types of reactors, distillation columns, combustion system and many more that are practicing unitability. There are many systems in the environment and social fields that are behaving naturally as unstable. Apart from this, there are some systems in the military and aviation areas, that are deliberately designed to be unstable to gain better maneuverability. It also helps in increasing the speed of command responses.

These types of systems are considered difficult to handled and more difficult to control than the stable systems. This is because of the designs and control engineers. To have safe control system, the extra attention is being payed.

There are several scientists that are working on the system stability and unitability. Like the classical condition that is explained to have thermodynamic stability of equilibrium (cf. Callen 1960). It is considered as a fundamental tool that can help in examining the stability of the equilibrium states, that is in a different branch of science. That is also in the form that requires maximum entropy for constant energy of an isolated system. It is directly based on second law of thermodynamics. For the particular form of material, the maximum entropy condition is too restrictive and also have limited applicability.

In this chapter, there are several important points, which are explaining the stability criteria in the thermodynamic system. The concept of thermodynamics of irreversible processes (cf. Meiksner and Reik, 1959; de Groot and Mazur, 1962) are being accepted, including the viable approaches. In case of solids, it is being explained that the entropy of the production in the transition that takes place between two nearby equilibrium state does not depend on the transition duration. This type of property refers to the rate, that is independent of dissipation.

A different but related topic is the stability of a quasi-static deformation process, which is dissipative, rate independent and is being introduced by the loading programs. There are several phenomena like bifurcation that can predict the critical instant in many cases. Beegle, Model, and Reid (1974b) in their recent discussion explained the use of Legendre transformation to establish the thermodynamic stability.

6.2. THERMODYNAMIC SYSTEM

Thermodynamic System is defined as a microscopic region or a space in the universe. It is a region in which one or more thermodynamic processes takes place. Everything which is external to a thermodynamic system called surrounding.

System and surrounding are separated by a defined border, that border is known as boundary. System, surrounding, and boundary together constitute and makes the universe.

6.3. DEFINING STABILITY

Stability, in terms of mathematical definition can be formulated as the ability to recover from perturbation. Stability can be either be defined as if the bounded input into the system produces bounded output from the system.

In terms of solution of different equation, function $f(x)$ is said to be stable. When any other solution of the equation starts out sufficiently close to it when the value of $x = 0$. It remains close to it so as to have increase in the value of x. if the difference reaches or approaches zero as x increases. This type of solution is called asymptotically stable. But if the solution does not have either of these properties, then this type of solution is called unstable.

In physical problems, it is important to have the stability in solution. This solution stability is important because if slight deviation from mathematical model is caused by the errors in measurement. The mathematical equations that are describing the problem, it will not accurately predict the future outcome.

Therefore, during prediction of the population growth it is considered on the difficulty, this is because of the fact that it is governed by the equation y

$= ax^{ce}$. Which is considered as an unstable solution of the equation $y' = ay$. Therefore, a slight error like in initial count c, or in the breeding rate a, can cause large errors during prediction.

6.4. TYPES OF THERMODYNAMIC SYSTEM

Thermodynamic system is classified into three types:

1. Open System;
2. Closed System; and
3. Isolated System.

1. Open System

Open System is a thermodynamic system. In this type, both mass and energy flows in and out of the system. This type of flow takes place across the boundary of the system. Example of engine is considered as an open system. In case of engine, when fuel is provided to the engine then it will produce a power which is given out. Thus, there is an exchange of mass as well as energy. Engine during this process also emits heat which is exchanged with the surroundings.

Another example of open system is boiling water. When water is boiled in an open vessel, then heat gets transfer with the exchange of mass in the form of steam between the vessel and the surrounding.

2. Closed System

A closed system is where the transfer of energy takes place. Inn this type of system, energy exchange takes place across its boundary with the surrounding. But in this case no mass transfer takes place. A closed system is considered as a fixed mass system.

Fluids like air or gas is being compresses in the piston and cylinder arrangement, these are the example of closed system. In this case, mass of the gas remains constant. But the gas is heated or cooled.

Another example of closed system is, when water is being heated in a closed vessel. Then the water gets heated and mass of water remains same. As water cannot escape outside the boundary of the system. It remains inside the vessel and maintain the mass of water.

3. Isolated System

Isolated system is when the system does not interact with its surrounding. This type of system does not allow the exchange or transfer of mass and energy across the boundaries. This time of system is considered as more restrictive.

In reality, it is considered that this type of isolated system does not exist. However, there are some systems that behaves like an isolated system for a period of time.

Taking an example of the piston and cylinder. When piston and cylinder is arranged with the fluid like air and gas, which is being compressed or expanded to insulate, then it becomes isolated system. In this case, there will not be transfer of mass nor energy.

Hot water, coffee or tea when kept in a thermos flask then this type of system is called a cold system. But when this material is poured in a cup, then this system become an open system.

6.5. THERMODYNAMIC EQUILIBRIUM

Thermodynamic equilibrium is a condition or state of the system. The property of which does not change with time and it can be only changed to another condition only at the time of expense on another system. In thermodynamic equilibrium, state with given pressure and temperature, the Gibbs free energy is smaller than that of any other state with the same pressure and temperature.

6.6. STABILITY CONDITIONS

When a particle moves in an external potential U and settles to a stable equilibrium with a minimum value of U. Instead of force U that gets vanish, this is a result of the loss of energy to friction processes. When potential energy is at the minimum, then stable equilibrium is said to occur. Equilibrium occurs for a thermodynamic system when, appropriate potential reaches to the extremum. For example, minimum value of entropy for an isolated system. When the requirement changes spontaneously, then this will lead to an increase in the entropy. This places important constraints on equilibrium response functions.

Consider a homogeneous system is at equilibrium. This can be characterized by intensive state functions (T, J, μ), and for extensive variables (E, x, N). Now let us imagine that the system is arbitrarily divided into two equal parts, where one part spontaneously transfers some amount of energy to other parts, that is in the form of work or heat.

Considering two subsystems, A, and B. They have initially same values for intensive variables. Meanwhile, their extensive coordinates satisfy

EA + EB = E, xA + xB = x and NA + NB = N

Energy exchange takes place between the two sub-system, after that the coordinates of A changes to

$$(E_A + \delta E, x_A + \delta x, N_A + \delta_N), \text{ and } (T_A + \delta T_A, J_A + \delta J_A, \mu_A + \delta\mu_A) \quad (1)$$

And coordinates of B changes to

$$(E_B - \delta E, x_B - \delta x, N_B - \delta N), \text{ and } (T_B + \delta T_B, J_B + \delta J_B, \mu_B + \delta\mu_B) \quad (2)$$

It is to be noted that system is maintained at a constant E, x, and N. Therefore, intensive variables are meant to be themselves functions of the extensive coordinates.

In reference to the first order in the variables of (E, x, N), then equation becomes:

$$\delta T_A = -\delta T_B \equiv \delta T, \delta J_A = -\delta J_B \equiv \delta J, \delta\mu_A = -\delta\mu_B \equiv \delta\mu \quad (3)$$

Using the entropy of the system the equation can be written as

$$S = S_A + S_B = \left(\frac{E_A}{T_A} - \frac{J_A}{T_A} \bullet x_A - \frac{\mu_A}{T_A} N_A\right) + \left(\frac{E_B}{T_B} - \frac{J_B}{T_B} \bullet x_B - \frac{\mu_B}{T_B} . N_B\right) \quad (4)$$

Assuming expanding the equation about the equilibrium point, second order of equation becomes

$$\delta S = \delta S_A + \delta S_B = \left[2\left(\delta \frac{1}{T_A}\right)\delta E_A - \delta\left(\frac{J_A}{T_A}\right).\delta x_A - \delta\left(\frac{\mu_A}{T_A}\right).\delta N_A\right] \quad (5)$$

Eq. (3) notes that the second order contribution of B is same as A. Eq. (5) can be rearranged as

$$\delta S = \frac{2}{T_A}\left[\delta T_A\left(\frac{\delta E_A - J_A.\delta x_A - \mu_A \delta N_A}{T_A}\right) + \delta J_A.\delta x_A + \delta\mu_A.\delta N_A\right]$$

$$= \frac{2}{T_A}[\delta T_A \delta S_A + \delta J_A \cdot \delta x_A + \delta\mu_A \cdot \delta N_A] \quad (6)$$

The condition that is used for stable equilibrium is, any kind of change that should lead to a decrease in entropy, and hence it is important to understand that

$$\delta T\, \delta S + \delta J \cdot \delta x + \delta\mu \cdot \delta N \geq 0. \quad (7)$$

In the above equation subscript, A is removed, so that the above condition can be applied to the whole system as well as to any part of the system. Above equation is obtained by the keeping the overall system at constant, that is E, x, and N. Since all the coordinates appears symmetrically in the above equation. Other sets of constraints will obtain the same result. Example of such result, variations in δT and δx with δN = 0. This will lead to:

$$\begin{cases} \delta S = \left.\dfrac{\delta S}{\delta T}\right|_x \delta T + \left.\dfrac{\delta S}{\delta x_i}\right|_T \delta x_i \\ \delta S = \left.\dfrac{\delta J_i}{\delta T}\right|_x \delta T + \left.\dfrac{\delta J_i}{\delta x_j}\right|_T \delta x_j \end{cases} \tag{9}$$

While substituting these variations into Eq. (7) gives

$$\left.\frac{\delta S}{\delta T}\right|_x (\delta T)^2 + \left.\frac{\delta J_i}{\delta x_j}\right|_T \delta x_i \delta x_j \geq 0 \tag{9}$$

It is to be noted that the cross terms proportional to $\delta T\ \delta x_i$ cancel due to the Maxwell relation. Eq. (9) is a quadratic from, so it must be positive for all choices of δT and δx.

Resulting constraints on the coefficient is independent of how the system was initially partitioned into subsystems A and B. this also represents the condition for stable equilibrium.

Considering that only δT is non-zero, Eq. (7) requires $\partial S/\partial T|_x \geq 0$, implying a positive heat capacity, since the equation derived as

$$C_x = \left.\frac{dQ}{dT}\right|_x = T\left.\frac{\delta S}{\delta T}\right|_x \geq 0$$

Considering one of the δx_i in Eq. (7) as non-zero, then the corresponding equation response function

$$\delta x_i / \delta j_i |_{T, x_j \neq i}$$

must be positive. However, all δx values may be chosen as non-zero in a more general requirement. The general requirement is matrix of coefficients

$$\delta J_i / \delta x_j |_T$$

must be positive definite. Matric can be considered as positive definite only if all the eigenvalues are positive. It is necessary that all the diagonal element of the matrix be positive. This leads to further constraints that can be between the response function. This includes chemical work for the gas. The appropriate matrix that can be used in this situation is

$$-\frac{\partial P}{\partial V}\bigg|_{T,N} \quad -\frac{\partial P}{\partial N}\bigg|_{T,V}$$
$$-\frac{\partial \mu}{\partial V}\bigg|_{T,N} \quad -\frac{\partial \mu}{\partial N}\bigg|_{T,N} \qquad (10)$$

Considering the positivity of the response functions $\kappa_{T,N} = -V^{-1}\,\partial V/\partial P|_{T,N}$, and $\partial N/\partial \mu|_{T,V}$, the determinant of the matrix must be positive, according to this equation derived as

$$\partial P/\partial N|_{T,V}\ \partial \mu/\partial V|_{T,N} - \partial P/\partial V|_{T,N}\ \partial \mu/\partial N|_{T,V} \geq 0. \qquad (11)$$

Another consequence of eq. (7) which is really interested, it relates to the critical point of a gas where $\partial P/\partial V\ |_{Tc,N} = 0$.

If the critical isotherm can be analytically expanded as

$$\delta P(T = Tc) = \partial P/\partial V\ |_{Tc,N}\ \delta V + 1/2\ \partial^2 P/\partial V^2|_{Tc,N}\ \delta V^2 + 1/6\ \partial^3 P/\ \partial V^3|_{Tc,N}\ \delta V^3 + \qquad (12)$$

the stability condition $-\delta P\ \delta V \geq 0$ implies that $\partial^2 P/\partial V^2|_{Tc,N}$ must be zero, if first derivative is found vanishing, then the third derivative is found to be negative.

from mean-field approximations to the isotherms (such as the van der Waals isotherms), this condition can obtain critical point of the gas. This usually does not justify making Taylor expansion around the critical point in Eq. (12). Even though the constraints $-\delta P\ \delta V \geq 0$ remains applicable.

6.7. THERMODYNAMICS AND STABILITY

1. If a thermodynamic system consists of lower potential energy. Then that system is considered as more stable.
2. Chemical process usually occurs because they are thermodynamically favorable. That is**DG** = -ve.
3. Thermodynamically favorable mean process from higher energy to lower energy or from less stable to more stable.
4. It is important to understand the relative stability of the molecule to predict relative reactivity of starting material and the relative yield of potential product.
5. Comparing experimentally measures or theoretical calculations stabilities, this can be determined.

6. For determining the relative stability, it is important to compare isomeric system.

6.7.1. Basics

- **DG** is the change in the free energy of the system,
- **DH** is the change in the enthalpy of the system,
- **DS** is the change in the entropy of the system,
- **T** is the absolute temperature in Kelvin,
- **R** is the universal gas constant, where value of R = 8.314 JK^{-1} mol^{-1} or 1.986 $calK^{-1}mol^{-1}$, and
- **K** is the equilibrium constant.

There are two very important fundamental thermodynamic equations that should be known and understanding that this relates to various thermodynamic terms:

DG = DH TDS

$$[\Delta G = -RT \ln k]$$

6.7.2. Heat of Reaction, DH_r^o

When in a particular reaction heat is released, then that heat is known as Heat of reaction. When a heat is released during this process, then the reaction is called exothermic reaction. In this heat is given out.

But when heat is absorbed during this process, then this type of reaction is known as endothermic reaction. In this process heat is taken in.

Example of endothermic reaction includes dissolving a salt. It doesn't have to be table salt, nor does the solvent need to be water. Some of the examples explaining best endothermic reaction

The reaction of barium hydroxide octahydrate crystals with dry ammonium chloride;

- Dissolving ammonium chloride in water;
- The reaction of thionyl chloride ($SOCl_2$) with cobalt (II) sulfate heptahydrate;
- Mixing water and ammonium nitrate;

- Mixing water with potassium chloride;
- Reacting ethanoic acid with sodium carbonate; and
- Photosynthesis, Chlorophyll is used to react carbon dioxide plus water plus energy to make glucose and oxygen.

The below examples can be written as chemical reactions in exothermic reaction, but are more generally considered to be endothermic or heat-absorbing processes:

- Melting ice cubes;
- Melting solid salts;
- Evaporating liquid water;
- Converting frost to water vapor (melting, boiling, and evaporation, in general, are endothermic processes);
- Making an anhydrous salt from a hydrate;
- Forming a cation from an atom in the gas phase;
- Splitting a gas molecule;
- Separating ion pairs;
- Cooking an egg; and
- Baking bread.

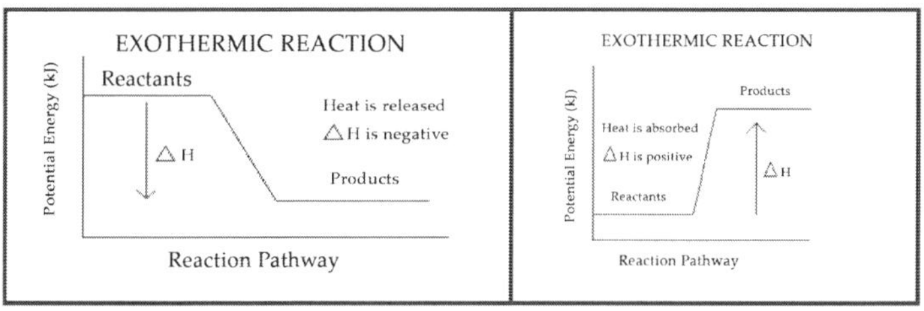

Figure 6.1. *Heat reaction.*

Source: Image by sites.prairiesouth.ca

There are certain types of reactions for making thermodynamic comparison:

6.7.3. Heat of Combustion, ΔH^0_c

Heat of combustion is defined as a heat released when one mole of a compound undergoes complete combustion of O_2. This is usually an exothermic process.

$2CO_2 + 3H_2O$ CH_3CH_3 (In presence of ΔH_c^0)

In the above example Exothermic reaction takes place. It should be noted that for heats of combustion, the organic compound is a starting material for the reaction.

6.7.4. Heat of Formation, ΔH_f^o

Heat of formulation is defined as the heat released if one mole of a compound were formed from its component elements in their standard state.

$CH_3CH_3 \Rightarrow 2C + 3H_2$ (in the presence of ΔHf^0)

Above example can be of endothermic reactor or can be of endothermic reaction. It should be noted that the organic compound is a product for the reaction for heat of formation.

6.7.5. Heat of Hydrogenation, ΔH_h^0

Heat of hydrogenation is the process in which heat is released upon the addition of H_2 to one mole of a compound to generate the corresponding result. Example if alkene or alkyne is used then the generation of alkane takes place.

In the below example, it is usually exothermic process. In the Below example for ethene to ethane, since 2 σ bonds are made from each π bond and each H-H σ bond. It should be noted that both starting materials and products are organic compounds for heat of hydrogenation.

$Ch_3CH_3 \Rightarrow CH_2 = CH_2 + H_2$

6.7.6. Hess's Law

Hess law is considered as very useful during investigating the thermodynamics of the reaction. Hess law can be expressed as "the enthalpy changes for a reaction (ΔH_r^o) that converts starting material to product. This works independent of the pathway.

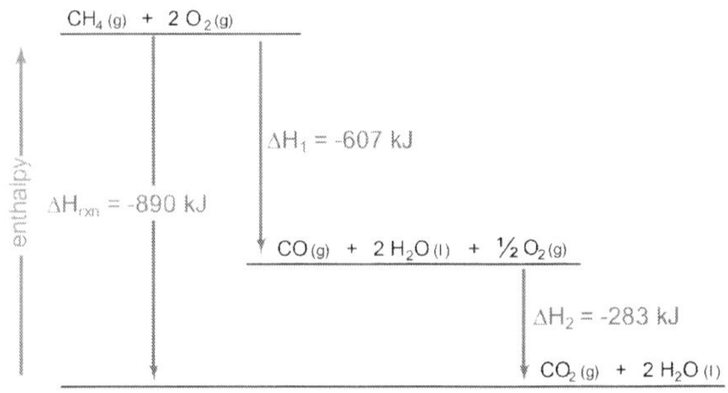

Figure 6.2. *Example of Hess law.*

Source: Image by ch301.cm.utexax.edu.

Hess law implies that: $\Delta Hrxn = \Delta H_1 + \Delta H_2 + \Delta H_3$

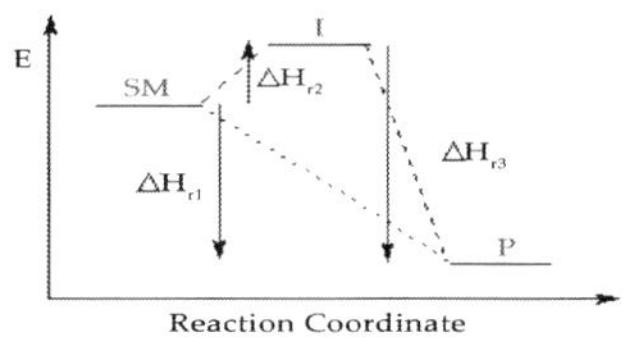

Hess's Law implies that: $\Delta H_{r1} = \Delta H_{r2} + \Delta H_{r3}$

Figure 6.3. *Hess law reaction.*

Source: Image by chem.ucalgary.ca.

For the reaction SM to P which can either proceed directly or via an intermediate, I, the overall enthalpy change for the conversion of SM to P must be the same.

i.e. $\Delta H_{SM-P^0} = \Delta H_{SM-I^0} + \Delta H_{I-P^0}$

6.8. SYSTEM STABILITY

6.8.1. Evolution of an Isolated System

As Entropy can be partially developed by the fact of Nature that "an isolated system can only evolve towards a unique state called equilibrium state," independent of its initial state. It is characterized by a particular internal distribution of conservative magnitudes.

That is, of maximum average uncertainty, i.e., maximum entropy. For a compressible system, its internal energy, U, and volume, V. The time evolution is dictated by the monotonous increase of entropy, S, constrained by the constancy of U and V. This is a statement of the Second Law of Thermodynamics, namely:

$$\frac{\partial S}{\partial t} \equiv \frac{\partial \sum S_k}{dt} > 0, \quad \frac{\delta U}{\partial t} \equiv \frac{\partial \sum U_k}{dt} = 0, \quad \frac{\partial V}{\partial t} = \frac{\partial \sum V_k}{dt} = 0 \text{ until} \lim_{\to\infty} \frac{\partial S}{\partial t} = 0 \tag{13}$$

At least for small departures, from equilibrium in an isolated system. The dynamics should be proportional to the unbalance force, the entropy function must be quadratic in the force:

(At least when $t \to \infty$)

$$\frac{\partial^2 S}{\partial t^2} < 0 \text{ until} \lim_{t\to\infty} \frac{\partial^2 S}{\partial t^2} = 0 \tag{14}$$

It's a fact that thermodynamic evolution is governed by a variational principle is a general behavior of nature. In Eq. (13), the trend towards a stationary value of entropy, under the energy and volume constraints.

This was first enunciated by Fermat in the 17th c. for Optics: "light rays travel from one point to another by the path of minimum time of travel," and later developed by Hamilton in the 20th c. for Mechanics: "a body moves in time-space from $(t_1\vec{r}_1)$ to $(t_2\vec{r}_2)$ by minimizing the action

$$A \equiv \int_{t_1}^{t_2} L(q_i, \dot{q}, t)dt = \min"$$

(where L is the Lagrangian function, with units of energy, q are coordinates and p momentum).

Similarly, in case of a mechanical system, thermodynamic systems can be stable or not. Considering a 1mJ spark can make to a hydrogen-oxygen mixture or the change in a hot pad when clicking the trigger.

6.8.2. Consequences of Equilibrium

In case of Entropy, it is found that the constraints on the distribution of conservative magnitudes is at equilibrium. With the help of Lagrange multipliers, it is found that a first consequence of a system being at equilibrium, is the existence of temperature and pressure. As common variables are shared by all subsystems k that works at equilibrium in absence of external fields:

$$\frac{1}{T} \equiv \left.\frac{\partial S_k}{\partial U_k}\right|_V = \left.\frac{\partial S}{\partial U}\right|_V, \frac{p}{T} \equiv \left.\frac{\partial S_k}{\partial V_k}\right|_U = \left.\frac{\partial S}{\partial V}\right|_U, \forall k \tag{15}$$

It is to be noticed that Eq. (15) only apply to the system at equilibrium, with contrary to Eqs. (13) and (14).

6.8.3. The Importance of Temperature as Positive

Temperature is defined by Eq. (14) can be only positive for a real system at equilibrium. This happens because of the fact of nature. This means that more energy a system has, the more quantum energy level become accessible.

Therefore, a larger entropy which means the number of possible distributions, and the temperature is that sensitive, i.e., $S(U)|_V$ can be only positive and increasing.

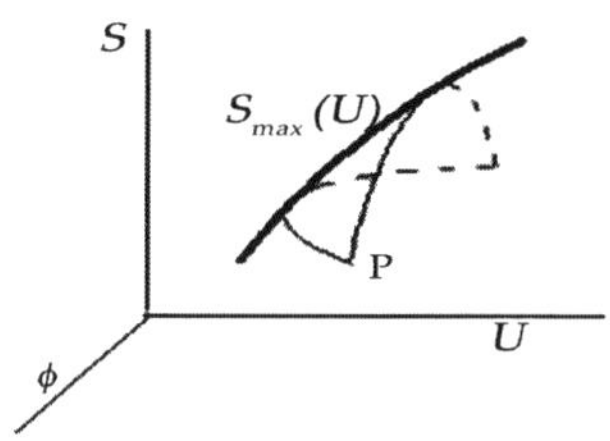

This figure shows the Entropy variation with internal energy, U (at constant volume), and with an unspecified non-equilibrium parameter, φ (e.g., the mean temperature gradient). Notice that from a non-equilibrium state P, the equilibrium state reached is different if the evolution is at U=constant (maximum entropy) or at S=constant (minimum energy).

The positiveness of $1/T=\partial S/\partial U|_V$ is implied in increasing monotony of the function $S(U)|_V$. Moreover, at constant volume, energy itself must be a monotonic increasing function of temperature, $\partial U/\partial T|_V>0$. Otherwise the system can be thermally unstable. Which means that energy would flow from colder to hotter parts and entropy would decrease. Thus, the function $S(U)|_V$ is monotonically increasing with monotonically decreasing slope, as represented in figure above.

The description above is based on quantum physics. But a mechanics reasoning serves the same goal: why temperature must be positive. The total energy of an isolated system, E (constant), can be composed as the sum of internal energy, U, and kinetic energy of the motion relative to its center of mass,

$E_k=\sum mivi^2/2$, i.e., $E=U+E_k=\text{const}$, and $dE/dt=0=dU/dt+dE_k/dt$.

But mechanical stability demands $dE_k/dt<0$ i.e., the expected conclusion explain that the motion tends to disappear and increase in the system thermal energy. Since $dS/dt>0$ for the evolution, it finally implies that $dU/dS>0$.

6.8.4. Extended Mechanical Consequences of Equilibrium

In case of Entropy, it is shown that the consequences of equilibrium can be labeled as follows:

- Thermal equilibrium: At equilibrium, temperature must be uniform in all subsystems. This is sometimes called Zero Law of Thermodynamics.
- Mechanical equilibrium: At equilibrium, the only motion possible for the subsystems is a solid body.
- Chemical equilibrium: At equilibrium, the chemical potential of each species varies proportionally to its molar mass and the energy of the applied field by being uniform in the absence of external fields.

The reasoning mechanism that is used above to show that temperature must be positive, this mechanism can be further extended to an isolated system with the total energy E that is constant, total angular momentum $\vec{L}$ (constant) and total linear momentum $\vec{P}$ (constant), in a constant gravitational field, that gives the equation as

$$E = U + \frac{1}{2}mv_{cm}^2 + \frac{1}{2}I\Omega^2 + mgz = U + \frac{\vec{P}^2}{m} + \frac{\vec{L}^2}{1} + mgz = \text{cont.} \quad (16)$$

with reference to traditional mechanical definitions, center of mass, $\vec{r}_{cm}$, inertia-averaged angular speed, $\vec{\Omega}$, mass-averaged velocity, $\vec{v}_{cm}$, moment of inertia relative to an arbitrary origin, *I*, angular momentum, $\vec{L}$ and linear momentum, $\vec{P}$:

$$r_{cm} = \frac{\sum m_i \vec{r}_i}{\sum m_i}, \vec{v}_{cm} \equiv \frac{\sum m_i \vec{v}_i}{\sum m_i}, I = \sum m_i (\vec{r}_i - \vec{r}_{cm})^2, \vec{\Omega} = \frac{\sum (\vec{r}_i - \vec{r}_{cm}) \times m_i (\vec{v}_i - \vec{v}_{cm})}{\sum m_i (\vec{r}_i - \vec{r}_{cm})^2} \quad (17)$$

$$\vec{P} = m_i \vec{v}_i = m\vec{v}_{cm}, \text{ and } \vec{L} = \sum \vec{r}_i \times m_i \vec{v}_i = I\vec{\Omega} + \vec{r}_{cm} \times \vec{P}$$

Deducing Eq. (17) from (16) and from $dU/dS>0$, at equilibrium. The thermal energy is said to be at maximum compatibility with the constraints. This is done by mechanical energy dissipation by friction.

With maximum compatibility with constrains, it also important that consequently the mass should be the highest. But this is constant in this non-relativistic analysis. With this the moment of inertia should also be the highest. This means that the deformable systems tend to rotate around the principal axis of inertia. The position in the gravity field that should be the lowest. Consequently, at equilibrium, the angular speed is aligned with the angular momentum.

6.8.5. Importance of Pressure to Be Positive

Pressure defined by equation (3), for a real system it can be positive only at equilibrium. As a system has more volume, the more quantum energy levels become accessible and thus the larger the entropy in terms of the number of possible distributions, and pressure is that sensitivity. Therefore, there is increasing monotony of the function $S(V)|_U$ implies the positiveness of $p/T=\partial S/\partial V|_U$.

Moreover, since volume has to be also a monotonic decreasing function of pressure at constant energy or at constant temperature, $\partial V/\partial p|_U$, it follows that the function $S(V)|_U$ is monotonically increasing with monotonically decreasing slope, as represented in the figure of temperature importance of being positive, for $S(U)|_V$. In fact, in the case of perfect gases, both functions are quite similar:

$$SS_{ref} = mc_v \ln(T/T_{ref}) + mR\ln(V/V_{ref}) = mc_v \ln(U/U_{ref}) + mR\ln(V/V_{ref}).$$

From other point of view, pressure can be only positive because if the tendency of entropy in isolated system increases, then this means that its volume would disappear dV.

6.8.6. Importance of Thermal Capacities That Have to Be Positive

Specific thermal capacity at constant volume, c_v, must be positive. Because thermal stability is already explained while analyzing the reason why temperature was positive. From its definition, and analyzing the evolution of the slope of $S(U)|_V$:

$$mc_v \equiv T\frac{\partial S}{\partial T}\bigg|_V = \frac{\partial U}{\partial T}\bigg|_V, \frac{\partial}{\partial U}\frac{\partial S}{\partial U}\bigg|_V = \frac{\partial(1/T)}{\partial U}\bigg|_V = \frac{-1\partial T}{T^2\partial U}\bigg|_V = \frac{-1}{mc_v T^2} < 0 \qquad (18)$$

At the time when energy increases the slope must decrease. it is derived from Eq. (18) that c_v>0.

A similar result can be obtained for thermal capacity at constant pressure $c_{p,\ i.e.,}$ $c_p > 0$. Although in other cases extrapolation is not granted. For an instance, the thermal capacity along the liquid-vapor saturation line can be negative.

6.8.7. Importance of Compressibilities Must Be Positive

Compressibility is the effect of pressure on the volume of a system. The two compressibility coefficients can be defined:

$$k \equiv \frac{-1}{V}\frac{\partial V}{\partial p}\bigg|_T \quad \text{and} \quad k_s \equiv \frac{-1}{V}\frac{\partial V}{\partial p}\bigg|_S \qquad (19)$$

Both the equations should be positive for mechanical stability. This mechanical stability is based on $\partial^2 S/\partial t^2 \partial t<0$ during evolution of the system. The explanation is like the above explanation for c_v. Equilibrium function $S(V)|_U$ is monotonic, with slope p/T monotonically decreasing with time, entirely like function $S(U)|_V$:

$$\frac{\partial}{\partial V}\frac{\partial S}{\partial V}\bigg|_U = \frac{\partial(p/T)}{\partial V}\bigg|_U = \frac{-1}{mk}\left(\frac{1}{vT} + \frac{(kp-\alpha T)^2}{C_v kT^2}\right) < 0 \qquad (20)$$

This implies that κ>0 (κ is sometimes labelled κ_T).

6.9. THERMODYNAMIC CONDITIONS FOR STABILITY OF EQUILIBRIUM

6.9.1. Compound Thermodynamic System

Considering a compound system, *A*, that consists of the material sample *M*, a loading device as a work source, and constant absolute temperature $\vec{T} > 0$ of a surrounding heat reservoir.

Accordingly, the Helmholtz free energy ϕ and entropy S of material sample M are treated as functions of $G = (q, T, \alpha)$

therefore, q stands for generalized displacement, α stands for the internal state variable and T stands for the uniform temperature, but this may differ from absolute temperature $\overline{T}$.

Therefore, loading device can be defined as a mechanical conservative system which contains potential energy Ω which is a time differentiable function of (**q**, λ), this supply work W^{ext} (note that there is no heat) to *M*, so equation becomes,

$$\dot{W}^{ext} = P \cdot \dot{q}, \quad P = P(q, \lambda) \equiv -\Omega_{,q}(q, \lambda), \qquad (21)$$

In this equation, λ behaves as a loading parameter? Cases where q is not constrained kinematically, there consideration are restricted. For simplicity, the entropy of the mechanical devices is neglected. This is done so, because to interpret Ω as the internal energy of the device. The constitutive rate of equation for the vector of generalized loads P can be written as following

$$\dot{P} = \dot{P}^* - K^* \cdot \dot{q}, \; \dot{P}^* = -\Omega,_{q\lambda} \dot{\lambda}, \; K^* = \Omega, qq^{\cdot} \qquad (22)$$

In this equation matrix K * describes the effect of deformation-sensitive loading. It should not be excluded from general investigations of material stability (cf. Hill, 1978; Petryk, 2000a, b).

The Kinetic energy K_A of the system *A* is assumed merely assumed to be non-negative. It is also assumed to vanish in equilibrium. Internal energy and entropy of the heat reservoir are denoted by E_h and S_h, respectively. Assuming, the system *A* is thermally isolated, so that the heat *Q* supplied to the material sample *M* from the reservoir satisfies

$$\delta Q = -\delta E_h = -T\delta S_h$$

In case any extra is heat supplied to *M* by a disturbing agency is also included in δQ.

The total (internal + kinetic) energy E_A and total entropy S_A of the compound system *A* are

$$E_A = U + \Omega + E_h + K_A, \; S_A = S + S_h \qquad (23)$$

with *U* as the internal energy of *M*.

If the value of loading parameter λ is fixed and external disturbance is absent, then the system A is regarded as fully isolated. This type of fundamental property motivates the selection of A for thermodynamic stability investigations. An equilibrium state of A is characterized and for convenience is identified by $g=g^0$ constant in time at fixed λ. The use of the first law gives equation:

$$\dot{E}_A = \dot{W} + \dot{\Omega} + \dot{K}_A. \quad (24)$$

From the second law of thermodynamics it is obtain:

$$\dot{S}_A = \dot{S} - \dot{Q}/\overline{T} \equiv \dot{S}^i \geq 0 \quad (25)$$

It is assumed and introduced, that the entropy production rate $\dot{S}^i$ consists of two parts, thermal $\dot{S}_h$ and intrinsic rate-independent $\dot{S}_{in}$,

$$\dot{S}^i = \dot{S}^{th} + \dot{S}^{in}, \dot{S}^{th} = \left(\frac{1}{T} - \frac{1}{\overline{T}}\right)\dot{Q}, \dot{S}^{in} = \frac{1}{T}\dot{D}. \quad (26)$$

Generally, other contributions to $(\dot{S}_i - \dot{S}_{in})$ might exist here, but might be due to viscous dissipation, which are neglected here.

This would not influence the extended condition for stability. Based on the second law, it is assumed in the following strengthened form: both $\dot{S}^{in}$ and the complementary part of $\dot{S}^i$ are separately non-negative

$$\dot{S}^{in} \geq 0, \ \dot{S}^i - \dot{S}^{in} \geq 0 \quad (27)$$

6.9.2. Classical Condition for Stability of Equilibrium

Let's consider that the prefix Δ which denotes an increment in some quantity in a process, perturbed in general, leading from an equilibrium state g^0 to a state g,

$$\Delta\psi = \psi(G) - \psi(G^0) \text{ or } \Delta\psi = \psi|_G - \psi|_{G^0} \quad (28)$$

for a state variable or a quantity, which also depends on the transition path between the two states, respectively. All increments are considered at a fixed loading parameter λ up to the end of §4. The classical condition for stability of an isolated system A of an equilibrium state g^0 can be written down in the form

$$\Delta E_A = \overline{T}\Delta S_A > 0 \text{ for every } G \neq G^0 \quad (29)$$

The equivalence follows from the possibility of manipulating the value of ΔE_A. This is done by exchanging extra heat between external sources and the heat reservoir, without affecting the value of the left-hand expression in inequality used in Eq. (29).

$$T = \overline{T} \text{ and } Q = P \text{ in } G^0, \tag{30}$$

The condition of thermal and mechanical equilibrium in *A* gives a standard result, respectively which follows as a consequence if inequality in Eq. (29) by equating to zero the first quasi-static variation of the left-hand expression in inequality (29) at fixed α. However, if α can vary then inequality (3.10) in an equilibrium state implies also (cf. §3c) that

$$\delta E_A - \overline{T}\delta S_A = -A(G^0).\delta\alpha \geq 0 \text{ in } G^0 \tag{31}$$

As discussed in §2, the thermodynamic forces *A*, do not vanish in an equilibrium state, moreover, $\dot{\alpha}$ compatible with *A* and corresponding to a positive intrinsic dissipation rate must be satisfying the equation with strict inequality. This contradicts inequality (in Eq. (31)), unless there is no rate-independent intrinsic dissipation.

It can also be concluded that the classical thermodynamic condition (explained in Eq. (28)) does not provide a satisfactory basis for investigations of stability in materials or systems with rate-independent dissipation. The existing attempts (cf. Bazˇant and Cedolin, 1991) refer, in effect, only to equivalent (which is essentially elastic) systems without an explicit definition of intrinsic dissipation of rate-independent type.

6.9.3. Extended Condition for Stability of Equilibrium

From conditions of system discussed above, it is concluded that the maximum entropy inequality is too restrictive for materials or systems with rate-independent dissipation. Its interpretation as a sufficiency condition for stability of equilibrium, its fulfillment would also exclude local material instabilities at the micro-level.

By adoption the point of view of local instabilities appearing at a lower scale represent the physical source of the rate-independent inelastic behavior of the macroscopic material sample. Because of which the reason for the over-sufficiency of the classical condition becomes clear.

The concept of rate independent dissipation can be complied to extend the appropriate classical condition for thermodynamic stability. This can be done (Petryk, 1995) by taking, the second law of thermodynamics not in its fundamental form (Eq. 24) but rather in its strengthened version (Eq. 26).

Specifically, it is the second inequality (Eq. 27) which replaces that (Eq. 24) in the usual argument.

Functional V can be defined by

$$\Delta V \equiv \Delta E_A - T\Delta S_A + \bar{T}\Delta S^{in}, \tag{32}$$

whereas,

$$\Delta S^{in} = \int_{t^0}^{t} \dot{S}^{in}\, dt = \int_{t^0}^{t} \frac{1}{T}\dot{D}\, dt \tag{33}$$

It is the path-dependent intrinsic production of entropy in a perturbed process leading from an equilibrium state $G^0 = G(t^0)$ to a current state G(t). The value of V varies continuously with respect to time t in any kind of process, since the energy and entropy does the same. Moreover, in a process free of actual (persistent) disturbances, which follows any earlier temporary disturbance, the equation

$$\dot{V} = \dot{E}_A - \bar{T}(\dot{S}^{i} - \dot{S}^{in}) \leq 0, \tag{34}$$

from Eq. (27) and the energy conservation law $\dot{E}_A = 0$ at fixed λ. This shows that ΔV is a natural candidate for the Lyapunov functional. The following conclusion can be gathered (Petryk, 1993b, 1995): If

$$\Delta V > 0 \text{ for every } G \neq G^0, \tag{35}$$

and for every path connecting g^0 with g, then the equilibrium state g^0 of an isolated system A is stable with respect to the distance $|\Delta V|$.

These conditions (Eq. 35) may be regarded as an extension of the classical conditions (Eq. 29) to systems with rate-independent dissipation.

Further, the discussion of condition (Eq. 35) and its consequences for stability in continuous systems can be found in the references quoted above.

On substituting Eq. (22) into (32), following equation is obtain

$$\Delta V = \Delta U - \bar{T}\Delta S + \bar{T}\Delta S^{in} + \Delta\Omega + \Delta K \tag{36}$$

The theory of thermodynamic stability can be developed by using the functional E defined by

$$\Delta E = \Delta U - \bar{T}\Delta S + \bar{T}\Delta S^{in} + \Delta\Omega \tag{37}$$

The kinetic energy K_A is non-negative and vanishes in equilibrium, therefore, from condition (Eq. 35), the basic corollary: Stability of an equilibrium state g^0 (with respect to the distance $|\Delta V|$ is ensured when

$$\Delta E > 0 \text{ for every } G \neq G^0, \quad (38)$$

for every path connecting g^0 with g. It is to be mentioned that condition (Eq. 35) implies $K_A < \Delta V$. So that the kinetic energy in a process may satisfy inequality (Eq. 34) that is bounded by the initial positive value of ΔV adopted as a measure of an initial disturbance strength.

For isothermal processes, Eq. (37) can be simplify further as

$$\Delta E = \Delta\phi + \Delta D + \Delta\Omega \; (T \equiv \bar{T}), \quad (39)$$

or equivalently, on using equation

$$\dot{\phi} = \dot{W} - S\dot{T} - \dot{D}.$$

, to

$$\Delta E = \Delta W + \Delta Q (T \equiv \bar{T}) \quad (40)$$

Under the above restriction to constant temperature, the quantity E can be identified (cf. Eq. (23)) with the internal energy of the compound system A. Isothermal stability conditions for inelastic solids, which can be expressed in terms of the quantities in Eq. (39) or (40), which have been previously based on the energy balance, cf. Hill (1958), Nguyen and Radenkovic (1976), Petryk (1982), and Nguyen (1993).

Their thermodynamic justification follows from the extended condition (38) by the limiting passage from a non-isothermal case (Petryk, 1995).

The present aim is to examine consequences of the condition (explained in Eq. (38) for the discretized material sample *M*. The primary task is to examine the sign of ΔE after infinitesimal deviations from g^0.

In view of the non-existence of the standard variation of ΔS^{in}, an infinitesimal increment denoted by a prefix δ or the associated rate denoted by a superimposed dot have the meaning of a one-sided directional variation or derivative, respectively.

A virtual one-sided variation of Eq. (37), on substituting equations of Helmholtz free energy function, dissipation function D, (20) and (25) can be written as follows

$$\delta E = (Q - P)\cdot\delta q + (T - \bar{T})\delta S + \frac{T}{T} D(G, \delta\alpha) = A\delta\alpha \quad (41)$$

As a condition necessary for fulfillment of Eq. (38), in place of the previous unacceptable condition (32), in an equilibrium state can help in obtaining the following equation

$$\delta E = D(G^0, \delta\alpha) - A(G^0) \cdot \delta\alpha \geq 0 \,\text{in}\, G^0 \quad (42)$$

The infinitesimal increments can be replaced by forward rates that can be defined by equation

$$\dot{\psi}(t) = \frac{\delta\psi}{\delta t}, \quad \delta\psi = \psi(t + \delta t) - \psi(t)$$

It is important to be noted that while the condition in Eq. (42) is necessary for the fulfillment of inequality (28), its status as a necessity condition for stability.

All $\dot{\alpha} \in L(g^0)$ compatible with the actual thermodynamic forces $A(g^0)$ satisfy Eq. (42) with the equality sign, comparing equation

$$D(G, \dot{\alpha}) = A(G) \cdot \dot{\alpha} \text{ for all } \dot{\alpha} \in L(G)$$

Any other $\dot{\alpha}$ requires an instantaneous finite perturbation of averaging thermodynamic forces A. It is an important question to understand the circumstances such perturbations may be regarded as physically realizable. Only those perturbations which satisfy Eq. (42) with the equality sign will be used and thus it will not require instantaneous finite perturbation of actual thermodynamic forces. Consequently, by this second-order term are to be analyzed.

6.9.4. Second-Order Stability Conditions

Stability analysis can be now continued within the framework of multimode inelasticity, where $\dot{\alpha}$ is expressible through equation in terms of $\dot{\gamma}_K$'s.

Their result can be compared with differently formulated conditions, that is second-order stability conditions, which are given by (cf. Hill, 1958, 1978; Nguyen, 1993, 2000; Petryk, 1993b, 1995).

Infinitesimal increments are henceforth replaced by forward rates equation.

Eq. (41) can be written as

$$\dot{E} = (Q - P) \cdot \dot{q} + (T - \bar{T})\dot{S} + \sum_k \left(\frac{\bar{T}}{T} \pi_k^c - \pi_k \right) \overset{\circ}{\gamma}_k \qquad (43)$$

Which can be applied in any kind of state. In particular, inequality (42) is satisfied automatically for all $\dot{\gamma}_k \geq 0$ if $f_K (g^0) \leq 0$.

Under no instantaneous finite perturbations of driving forces πK, all the conditions in equation

$$\overset{\circ}{\gamma}_K \geq 0,\ f_K \leq 0,\ f_K \dot{\gamma}_k = 0,$$

are satisfied in the examined equilibrium state g^0. Denote by $P^0 = P(G^0)$ the set of indices of potentially active mechanisms in the state g^0, i.e., $K \in P^0 \Leftrightarrow f_K(G^0) = 0$, and by [ψ] a column matrix of components ψ_K with $K \in P^0$.

Relations like $\Psi \geq 0$ are read component wise. Then, the set of all $[\dot{\gamma}] \geq [0]$ can be identified with the set of virtual modes of inelastic deformation realizable under arbitrary small perturbations of driving forces π_K.

The required perturbations of π_K are initially zero in g^0 and their magnitude grows in time with a finite rate if the consistency conditions equation is violated in the state G^0. From Eq. (43) it follows that $\dot{E} = 0$ for every set $\left(\dot{q}, \dot{T}, \left[\dot{\gamma}\right]\right)$ in an equilibrium state g^0.

The further step is to examine the sign of the forward second-order rate of ΔE (with respect to an appropriate time scale) calculated as the rate of Eq. (43). On substituting next the equilibrium conditions and rearranging, this yields[2]

$$\ddot{E} = \left(\dot{Q} - \dot{P}\right) \cdot \dot{q} + \dot{T}(\dot{S} - \dot{S}^{in}) - \sum_{K \in P^0)} \dot{f}_K \dot{\gamma}_K \ \text{ in } G^0 \qquad (44)$$

The second-order condition for stability of equilibrium, corresponding to the basic condition (28), is expressed in the following equation

$$\ddot{E} > 0 \text{ for every non} - \text{zero}(\dot{q}, \dot{T}, [\dot{\gamma}] \geq [0]) \text{ in } G^0 \tag{45}$$

In general, the condition of the Eq. (45) is not enough for the stability. But it is important for the directional stability of an equilibrium state g^0. Most of the straightforward forms of instability are being excluded. Which can be named as, with the accuracy of the first order, spontaneous departure from equilibrium on a direct path along which all macroscopic state variables vary proportionally with respect to some distance measure.

Physical meaning of instability cannot be shown in the violation of inequality that is explained in Eq. (45). According to this, the inequality that is sown in Eq. (45) are interpreted merely as a sufficiency condition that can show the directional stability of a general or specified type, respectively.

From inequality (45) there are several more specialized conditions that can be all formulated in g^0:

For thermal stability

$$(\dot{q} = 0, [\gamma] = [0]) - \phi_{,TT} > 0, \tag{46}$$

i.e., the classical condition of positiveness of the specific heat at constant deformation.

For elastic stability $(\dot{T} = 0, [\gamma] = [0])$

$$\dot{q}(K^e + K^*) \cdot \dot{q} > 0 \text{ for every } \dot{q} \neq 0, \tag{47}$$

The classical condition of positive definiteness of the second variation of the total potential energy of the system basically coincides with this condition. For thermoelastic stability $\left[\dot{\gamma}\right] = [0]$ the conditions (46) and (47) jointly, and no other condition is generated. For intrinsic stability $(\dot{T} = 0, \dot{q} = 0)$

$$\sum_{K, L \in P^0} g_{KL} \gamma_K \gamma_L > 0 \text{ for every non} - \text{zero} \left[\gamma\right] \geq [0], \tag{48}$$

i.e., stability against spontaneous internal structural rearrangement, that is in the material sample at fixed macroscopic deformation and temperature. As a thermodynamic condition for material stability, Eq. (48) was derived first in (Petryk 1993b), although that inequality appeared earlier in the literature in a different context, cf. Havner (1992).

Henceforth, there is restrict attention to isothermal stability, assuming that $T \equiv \bar{T}$

On substituting equations

$$\dot{Q} = K^e \cdot \dot{q} - k\dot{T} - \sum_K N_K^P \overset{\circ}{\gamma}_K, \dot{f}_K = N_K \cdot \dot{q} + k_K \dot{T} - \sum_L gKL \overset{\circ}{\gamma}_L$$

and (21), the condition (25) reduces to the following second-order condition for isothermal stability of equilibrium of the material sample under deformation-sensitive loading

$$\dot{q}(K^e + K^*) \cdot \dot{q} - \dot{q} \cdot \sum_{K \in P^0} (N_K^P + N_K) \overset{\circ}{\gamma}_K + \sum_{K,L \in P^0} gKL \overset{\circ}{\gamma}_K \overset{\circ}{\gamma}_L > 0 \tag{49}$$

for every non-zero $(\dot{q}, \left[\dot{\gamma}\right] \geq [0])$.

The above condition includes conditions (47) and (48), as immediate consequences. It is important to note that the stability condition (49) and its consequences that follow, although derived from thermodynamic considerations, eventually involve only those constitutive parameters that appear in the phenomenological constitutive rate equations

$$\dot{Q} = K^e \cdot \dot{q} - k\dot{T} - \sum_K N_K^P \overset{\circ}{\gamma}_K, \quad \dot{f}_K = N_K \cdot \dot{q} + k_K \dot{T} - \sum_L g_{KL} \overset{\circ}{\gamma}_L.$$

for the material and (21) for the loads.

Applications of condition (49) that do not require specification of internal state variables or thermodynamic forces. The specific form (49) of the stability condition appears as new. It should be assumed that the condition (47) of elastic stability is satisfied.

This allows the rate of generalized displacements to be eliminated from the stability condition (cf. Nguyen 1984). Then the left-hand expression in condition (49) for given $\left[\dot{\gamma}\right]$ reaches a minimum with respect to q at

$$\dot{q}^m = \frac{1}{2} G^* \cdot \sum_{K \in P^0} (N_K^P + N_K) \overset{\circ}{\gamma}_K, \ G^* \equiv (K^e + K^*)^{-1} \tag{50}$$

On substituting Eq. (49), the following condition is obtained

$$\sum_{K,L\in P^0} (g_{KL} - \frac{1}{4}(N_K^p + N_K)\cdot G^* \cdot (N_L^p + N_L)\overset{\circ}{\gamma}_K \overset{\circ}{\gamma}_L) > 0 \qquad (51)$$

for every non-zero $\left[\overset{\circ}{\gamma}\right] \geq 0$

Clearly, the conditions (47) and (51) considered jointly are equivalent to condition (49). The condition (51) is more restrictive than condition (48) on account of the positive definiteness of g^*. Which implied by inequality (47).

The condition (51) ensures directional stability of equilibrium, this is for the most unfavorable combination of disturbances both at the macro-level (associated with $\dot{Q} \neq \dot{P}$) and at the micro-level (associated with $\dot{f}_k \dot{\gamma}_k \neq 0$ It may be thus over enough for directional stability.

Two less restrictive versions of inequality (51) are as follows. First, if external quasi-static equilibrium remains unperturbed then, from equations above and (21),

$$\dot{Q} = \dot{P} \Leftrightarrow \dot{q} = G^* \cdot \sum_K N_K^p \overset{\circ}{\gamma}_K \qquad (52)$$

On substituting Eq. (52), the condition (49) reduces to the following condition for internal stability under deformation-sensitive loading

$$-\sum_{K\in p^0} \dot{f}_K \overset{\circ}{\gamma}_K = \sum_{K,L\in p^0} g_{KL}^* \overset{\circ}{\gamma}_K \overset{\circ}{\gamma}_L > 0 \text{ for every non-zero}[\overset{\circ}{\gamma}_L] \geq [0] \qquad (53)$$

where,

$$g_{KL}^* = g_{KL} - N_K \cdot (K^e + K^*)^{-1} \cdot N_L^p \qquad (54)$$

The above equation is another condition for stability against spontaneous internal structural rearrangements in the material sample. Condition (54) is explicitly dependent on the constraint's stiffness *K** which contrasts with condition (47) applicable under rigid boundary constraints for *M*.

It should also be noted that, if the additional assumption of normality ($N^p_K = N_K$) is introduced then the conditions (51) and (53) coincide. Which means that a macroscopically quasi-static deformation mode in Eq. (52) is then most critical for internal stability.

Second, assuming that while there are no internal disturbances that would perturb directly the values of driving forces π_K, when alternatively, that Eq. (52) is perturbed. Then the consistency condition $\dot{f}_K \leq, \dot{f}_K \dot{\lambda}_k = 0$ for every $K \in P \equiv \{K : f_K = 0\}$ must be satisfied, which for each given $\dot{q}$ leads to the linear complementarity problem for unknown $[\mathring{\gamma}]$.

Let's suppose that the inequality (47) holds, which ensures existence of a solution $[\mathring{\gamma}]$ to that problem for every $\dot{q}$ (cf. Cottle et al., 1992). Then, Q in each state is expressible in terms of Q, although not necessarily unique. For those $\dot{Q}$, the condition (49) reduces now to

$$\dot{Q} \cdot \dot{q} > -\dot{q} \cdot K^* \cdot \dot{q} \text{ for every } \dot{q} \neq 0 \qquad (55)$$

For dead loading (K* = 0), the familiar condition $\dot{Q} \cdot \dot{q} > 0$ is recovered, which corresponds to Hill's (1958) integral condition for stability in a continuum. Conditions in the form (55) and (54) for material stability at locally uniform strain, were given by Hill (1978).

If all principal minors of the matrix (g_{KL}) with $K, L \in P^0$ are positive, then the solution g 8 ½ to the above-mentioned linear complementarity problem is unique; cf. (Cottle et al., 1992) for the mathematical theory and (Maier 1969; Petryk 2000a) for its applications to inelastic materials. Then $\dot{Q}$ in each state becomes a continuous and piecewise linear function of $\dot{q}$, and inequality (54) can be written down as follows

$$\dot{q} \cdot (\boldsymbol{K}(\dot{q}) + \boldsymbol{K}^*) \cdot \dot{q} > 0 \quad \text{for every } \dot{q} \neq 0, \quad \boldsymbol{K} = \frac{\partial \dot{Q}}{\partial \dot{q}}.$$

If $(\dot{Q}, \dot{q})$ is not differentiable at some q then inequality (55) is to be understood in the limit sense as this q is approached from any side.

K is identified as the tangent stiffness matrix for the material sample, dependent on the direction of $\dot{q}$ in a piecewise-constant manner. Of course, inequality (55) is different from (and less restrictive than) the requirement of positive definiteness of every well-defined matrix (K ($\dot{q}$) + K*).

6.10. CONCLUSION

It is important to maintain the stability of a system to ensure the correct working of the system. There are some systems that are unstable due to their requirement and nature. There are different approaches used to have system stability. Several researches have done to calculate the way out to have stability within the system. To have better understanding about the thermodynamic system stability it is necessary to understand the basic of thermodynamic system and to understand the various conditions thatare supporting system stability.

The minimum value of thermodynamic potential can characterize the state of equilibrium. That state of equilibrium corresponds to the variables that are independent of the conditions of the experiments that takes place within the system. For example, when some of the criterial like entropy, volume, number of moles, etc., of the component are independent variables. In this case the system internal energy must be minimum if the system is to be in thermal equilibrium.

It is important to understand the different criteria's that are used to maintain the stability of the thermodynamic system.

REFERENCES

1. Bright Hub Engineering, (n.d.). *What is a Thermodynamic System? Types of Thermodynamic Systems*. [online] Available at: https://www.brighthubengineering.com/thermodynamics/3733-what-is-a-thermodynamic-system/ (accessed on 11 March 2020).
2. Chem.ucalgary.ca. (n.d.). *Ch 2: Thermodynamics and Stability*. [online] Available at: http://www.chem.ucalgary.ca/courses/351/Carey5th/Ch02/ch2–7.html (accessed on 11 March 2020).
3. Chemistry LibreTexts, (2019). *17.14: Heat of Combustion*. [online] Available at: https://chem.libretexts.org/Bookshelves/Introductory_Chemistry/Book%3A_Introductory_Chemistry_(CK-12)/17%3A_Thermochemistry/17.14%3A_Heat_of_Combustion (accessed on 11 March 2020).
4. Encyclopedia Britannica, (n.d.). *Stability | Solution of Equations*. [online] Available at: https://www.britannica.com/science/stability-solution-of-equations (accessed on 11 March 2020).
5. Gazdoš, F., (2014). *Introducing a New Tool for Studying Unstable Systems*. [online] Available at: https://link.springer.com/article/10.1007/s11633–014–0844-z (accessed on 11 March 2020).
6. Heidemann, R., (1975). *The Criteria for Thermodynamic Stability*. [ebook] Canada. Available at: https://www.academia.edu/27792771/The_criteria_for_thermodynamic_stability (accessed on 11 March 2020).
7. Helmenstine, A., (2019). *Examples of Endothermic Reactions*. [online] ThoughtCo. Available at: https://www.thoughtco.com/endothermic-reaction-examples-608179 (accessed on 11 March 2020).
8. Mishra, P., (2016). *Thermodynamic System – Types of Thermodynamic System – Mechanical Booster*. [online] Mechanical Booster. Available at: https://www.mechanicalbooster.com/2016/04/thermodynamic-system-types-of-thermodynamic-system.html (accessed on 11 March 2020).
9. Sankararaj, S., (n.d.). *Thermodynamic System and its Types « Mechteacher.com*. [online] Mechteacher.com. Available at: http://mechteacher.com/thermodynamic-system/ (accessed on 11 March 2020).
10. Stability Conditions, (n.d.). [ebook] Available at: http://web.mit.edu/8.333/www/lectures/lec4.pdf (accessed on 11 March 2020).

11. System Stability, (n.d.). [ebook] Available at: http://webserver.dmt.upm.es/~isidoro/bk3/c04/System%20stability.pdf (accessed on 11 March 2020).
12. TheFreeDictionary.com. (n.d.). Thermodynamic Stability. [online] Available at: https://encyclopedia2.thefreedictionary.com/Thermodynamic+Stability (accessed on 11 March 2020).
13. Trans., P., & Soc, A, R., (2005). Thermodynamic Conditions for Stability in Materials With Rate-Independent Dissipation. [ebook] Poland. Available at: http://www.ippt.pan.pl/Repository/o618.pdf (accessed on 11 March 2020).
14. Umantsev, A. (1991). Thermodynamic Stability of Phases and Transition Kinetics Under Adiabatic Conditions. [ebook] Illinois. Available at: https://pdfs.semanticscholar.org/eca6/8040c7fe48cfe85ca25afd2a9466cf3ddc18.pdf (accessed on 11 March 2020).

CHAPTER 7

Computational Thermodynamics and Application

CONTENTS

In the chapter computational thermodynamics and its application, the role of computational thermodynamics and its important application in various aspects are discussed. The chapter explains the meaning of computational thermodynamics, its recent developments, and future potential and prospects. The chapter also explain the computational thermodynamics and informational technologies in the field of computer design of materials.

In addition, the chapter also highlighted two methods for Investigations of phase diagrams and thermodynamic properties. The chapter also discuss the meaning of CALPHAD method and its use in computational thermodynamics.

The chapter also describe the application of computational thermodynamics to steelmaking and processing. The chapter explain some interrelated terms that are computational thermodynamics, computational kinetics, and materials design. In the end, the chapter explain the integrated computational materials engineering (ICME) and MGI.

7.1. INTRODUCTION

Computational thermodynamics, also known as CALPHAD method when created in the year 1950, aimed at interlinking phase diagrams with thermo chemistry by the help of computational techniques. Its further advances toward kinetic simulations accompanied with thermodynamic calculations, for instance, computational kinetics, including precipitation simulation, diffusion-controlled phase transformation, and phase- field simulation.

> Computational thermodynamics is the use of computers to simulate thermodynamic problems specific to materials science, particularly used in the construction of phase diagrams.

In the meantime, thermodynamic, mobility, and physical property databases for multi-component and multi-phase materials, served as basic knowledge for materials design, are critically evaluated by CALPHAD approach combining key experiments, first-principles calculations, statistic methods, and empirical theories.

Computational thermodynamics can be defined as a phenomenological scientific discipline that enables materials scientists and metallurgical engineers to calculate and assess phase diagrams and to numerically study and simulate phase transformations and phase equilibria. The properties of thermodynamic as a function of temperature and composition can also be calculated.

The supreme practical value of computational thermodynamics refers to its skill and capability to permit users to model and numerically solve—with the help of using specialized and standardized computer software such as DICTRA, Thermo-Calc, FactSage, JMatPro, Pandat, and MTDATA—a large number of diverse and very multifaceted engineering problems related to materials.

Such a computational approach creates its good image in terms of most efficient and cost-effective way of studying and designing complex, modern alloys. In some cases, the strategy or approach on the basis of computational thermodynamics may itself yields good results trustworthy enough to be directly used.

The experimental approach, which is the traditional method of assessing the thermodynamic properties of binary and multicomponent systems and alloys, and of determining phase diagrams—is, needless to say, very time consuming and highly expensive.

The numerical techniques which is mostly adopted by the engineers of computational thermodynamics software can be categorized into two segments.Gibbs Energy Minimization technique, which helps in calculating and assessing the value of global minimum of the Gibbs energy for heterogeneous system.

Solving and finding solution to a phase equilibrium equations, in which it is assumed that the chemical potentials for the components of various phases in the system are equal.

7.2. COMPUTATIONAL THERMODYNAMICS: RECENT DEVELOPMENTS AND FUTURE POTENTIAL AND PROSPECTS

7.2.1. Motivation

Modern technology is usually associated with the computer-based development of novel, customized created alloys with intricate composition as well as competing structural phases. Some of the key objective or concern for the material scientists are, for example, the adaptive transformation of crystal structures under mechanical load, size control of precipitates, the stabilization, and the creation of alloys with special properties such as magnetic, electrical or thermal properties.

In majority of the instances, there is a need to have comprehensive understanding about the temperature dependent phase stabilities and the progress of thermodynamic quantities during the phase of transformations. Therefore, thermodynamic computer simulations, in addition with the extensive databases materials, are gradually becoming more important for scientists as well as for the materials producers. The precision of thermodynamic anticipation is prerequisite for computational development of new alloys. Therefore, the accuracy and reliability of the thermodynamic databases (TD) must be as highest as possible. Since all of these materials are the result of the elements in the periodic table, it is thus essential to have an understanding about the unaries for subsequent accurate modeling of multi-component systems.

Thus, in a way to work in a systematic manner with respect to the new and improved materials, it is essential that different development and research activities use the same set of recommended data for these unaries, resulting in precisely comparing the results. For the desired generalization to multi-component systems, it is further significant that all unaries are described in such a way to make comparison possible, for instance, using a similar set of functions.

7.2.2. CALPHAD Method

One example of the achievement of this strategy is documented by the SGTE (Scientific Group Thermo data Europe) unary database, which is the element of the CALPHAD (calculation of phase Diagrams) method for predicting the properties of metastable and stable phase equilibria in multi-component systems. The existing assumption behind this is that "lattice stabilities," which was first introduced by Kaufman and was primarily applied to relative stabilities, entropy, and enthalpy of transformation.

Empirical fashion mathematical models are designed in order to define these properties primarily in relation with polynomials as function of pressure, temperature, and for multi-component systems that is also a function of composition. The constraints of these models for systems with mainly fewer than four components are assessed using available experimental data. For the account of a multi-component system, the mathematical models of the unaries, binaries as well as ternaries are associated with the databases and the results from Gibbs energy minimization calculations obtained from such a database are then authenticated by engaging in evaluation with the experimental data of systems having greater than three components.

The CALPHAD method is presently the only available effective method that can work with precise accuracy for practical applications in multi-component as well as in the multi-phase systems. As a result of this, the method has gained popularity and become essential everyday tool in many industries, including some of the biggest largest (and conceived to be) most conservative, such as steelmaking industry that have high contribution in the GDP of a country.

For example, the method is seamlessly used to pretend the thermodynamic stability of precipitate phases, which is of high importance for the mechanical properties of steels. One of the other benefits of using the CALPHAD methodology is that its application is not only limited to thermodynamics but can also be applied to the description of any phase-based property.

For instance, for diffusion simulations, thermodynamics is united with the portrayal of mobility's to achieve diffusion coefficients. For the explanation of the composition dependence of mobility's, the similar formalisms can be used for thermodynamics. Furthermore, CALPHAD software has also been integrated with simulation tools like microstructure evolution software.

This extensive application is one of the primary reasons that why the CALPHAD method was recognized in the report of the U.S. National Research Council on Integrated Computational Materials Engineering (ICME) as the most imperative tool for ICME with the purpose to ensure the development of widespread capabilities in the future.

The primary function of CALPHAD in ICME makes it an essential element of the Materials Genome Initiative (MGI) [by providing support of relevant data that can be used in the development of new, advanced materials.

7.3. COMPUTATIONAL THERMODYNAMICS AND INFORMATIONAL TECHNOLOGIES IN THE FIELD OF COMPUTER DESIGN OF MATERIALS

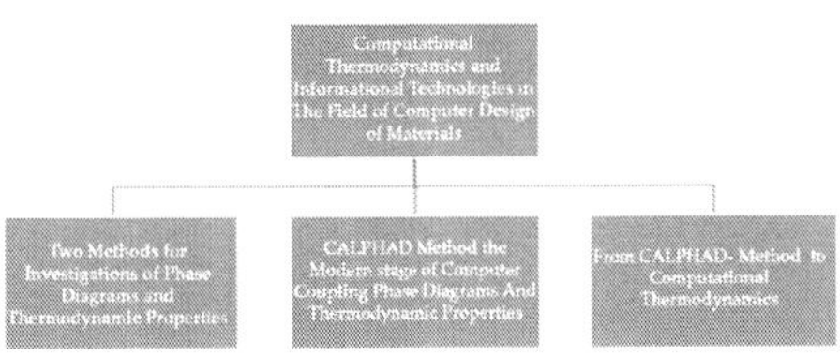

Figure 7.1. *Computational thermodynamics and informational technologies in the field of computer design of materials.*

7.3.1. Two Methods for Investigations of Phase Diagrams and Thermodynamic Properties

The phase diagrams are basically the pictorial representation of the physical state as function of pressure. temperature, and concentrations of the components.

There are generally two ways in phase diagrams investigations and in thermodynamic properties of multi-component systems, as explained by the materials scientists having experience in the field of physics and chemistry: by theoretical and experimental methods.

The experimental determination of thermodynamic properties and phase diagrams of binary and multicomponent systems is length, and thus time consuming and expensive.

Theoretical calculations of thermodynamic properties and phase diagrams makes the following tasks easy:

- It became possible to find a link or connection between experimental information on phase equilibria and thermochemistry;
- It allows to minimize the total amount and time spent in experimental investigations for formation of new phase diagrams for ternary and for more multi component system; and
- Allowing to anticipate the thermodynamic properties as function of temperature and composition, where there is no information about the experimental data.

7.3.1.1. Basic Stages of Historical Development of Interconnection between Thermodynamics and Phase Diagrams

The theoretical foundation and geometrical link or correlation between phase diagrams and thermodynamic properties was determined more than 100 years back ago by J.W. Gibbs. From theoretical perspective, the tie-line of the T-x phase diagram in case of binary closed system is the anticipation of the common tangent line to curves for Gibbs energy.

After J.J.Van Laar, B.Ja.Pines, J.Bekker, and D.S.Kamenetskaya done the theoretical investigations of hypothetical phase diagrams, it was the Larry Kaufman that started with his pioneering work on lattice consistency of pure elements (1959) for as computerize estimation of phase diagrams, published in 1970 with Bernstein in book "Computer Calculations of Phase Diagrams."

After the first issue of Journal CALPHAD, the International Journal "Bulletin of Phase Diagrams" was published in the year 1980, later changed its name to Journal of Phase Equilibria, which at the starting of its journal included papers on assessment of experimental data of phase diagrams and later it started covering papers dealing with calculational and experimental methods in examination of phase diagrams.

In the past few years, numerous international journals (such as Metall. Trans., J. Alloys Compounds, Journal of Nuclear Materials, Z. Metallkunde, Phys.Rev., Journal of Mining and Metallurgy and various others) include papers on assessment of thermodynamic properties of phase diagrams and different phases for binary and ternary systems.

7.3.2. CALPHAD Method the Modern Stage of Computer Coupling Phase Diagrams and Thermodynamic Properties

The strategy of the CALPHAD method include the next steps:

- The assortment and evaluation of experimental data on thermodynamic properties and phase equilibria of alloys for binary systems;
- The recognition of computer coupling and measurement of phase diagrams and thermodynamic properties binary systems;
- The planning or evaluation and getting of experimental data on ternary alloys; and
- Initiating necessary changes in the measurement of phase diagrams as well as in the thermodynamic properties of ternary systems or adjusted values of ternary interaction parameters for diverse phases in three component systems and etc.

The primary objective of CALPHAD – project is:

- The formation of data base for stable thermodynamic properties and metastable phases of pure elements, for example;
- The construction of stability in thermodynamic data bases of model description for variety of phases (melts, disordered solid solutions, chemical compounds, stoichiometric compounds, with including the homogeneity fields) as well binary as ternary systems, for example; and
- The development general software and algorithms for assessment and adjustment phase diagrams and thermodynamic properties in case of multi-component systems.

7.3.3. From CALPHAD-Method to Computational Thermodynamics

But the CALPHAD method up today has restrictions:

- The primary objective is to create or find connection between thermochemistry and phase diagrams only;
- It not only carries predictions and coupling physical properties (parameters chemical compounds and crystal structure solutions, temperature expansion coefficient, elastic modulus, etc.) in addition with the phase diagrams pure properties, and thermodynamic properties, different phases of ternary and binary systems;
- It use very basic phenomenological models, that have no base with the theoretical segment;
- It also makes uses of advance software packages (Thermo-Calc, 1985), Lukas programs (1977), Chemsage (1990, 1995), F*A*C*T, MT-DATA.

Within the past 25–30 years, it was observed that the application for the calculation of the thermodynamic of phase diagrams for both binary and multi-component systems has been increased. The computer programs which was formed in different research centers all over the world are of very high importance.

The quantitative methods that these programs cover can be categorized into two different segments (Counsel, 1971), the Gibbs energy minimum is sough for heterogeneous system. In the studies on the second class, the various set of phase equilibrium equations (determining the equivalence of chemical potentials for the elements of various phases) is resolved (Kaufman, 1970).

In this scenario, either the Nelder – Mead modified simplex method or Newton-Raphson iterative method is used to lessen the goal function stating the sum of residual of phase equilibrium equations. One major restraint of using the latter methods is the impossibility of assuring achievement of the universal least of the Gibbs energy for heterogeneous system. it can be assumed that the calculated phase diagrams can both be metastable and stable.

A major constrains of using the former method is the prerequisite of selecting a successful initial estimation for initiating the iterative process, for which there is no guarantee that the iterations will result in getting the wanted solution. Thus, it can be said that the calculating methods for both first and second class, which is considered as the first-generation software, render it chiefly impossible to generate autonomic computer programs for measurement of thermodynamic properties and phase diagrams of multicomponent alloys of systems with three or more than three components.

Therefore, in frame computational thermodynamics were developed well-founded (from the quantitative perspective) method for calculations of thermodynamic properties and phase diagrams of multi-component systems. Thus, the current computational thermodynamics include four basic "component": (1) logical thermodynamics of phase equilibria, (2) physical models, required for explanation of thermodynamic properties of different types phases (melts, disordered solutions, chemical compounds with or without homogeneity fields and etc.), (3) mathematical-founded software, and (4) experimental data on thermodynamic properties and phase equilibria of alloys.

7.3.3.1. Binary "Interactions" Analytical Thermodynamics and Experimental Data

Applications of some outcomes of analytical thermodynamic to evaluation of experimental data make it possible to choose the accurate data in between the working of experimental data, attained by various scientists. This result also allows the scientists to conduct an assessment for compositions at the temperature of non-variant equilibria for some binary systems.

One example of application of experimental data and analytical thermodynamics on phase equilibria are the outcome of measuring the temperature FCC->L virtual phase transformation for metastable FCC – phase Cr.

The overview of "equality of the areas Maxwell rule" in coordinates $\partial G/\partial x - x$, where G is the molar Gibbs energy of phases as function of composition, while x is a composition of second chemical component in binary system that make it easy to form an autonomic and original algorithms program for calculations of binary systems, covering up to 10 disordered solutions and up to 10 chemical stoichiometric compositions compounds.

Its simplification from binary to ternary systems make it easy to form an original algorithm and computer program for assessment of isothermal sections in the two-phase fields of ternary systems. Within the context of vector approach to the thermodynamics of multicomponent systems, the computer program and algorithm for calculations of several properties of thermodynamics (chemical potentials of components, specific heat, mole fractions of phases) as function of temperature for two-phase alloys with persistent composition.

Physical models criteria and analytical thermodynamics of concordance for outcomes of ab-initio calculations of variance energies between several phases at zero K and the thermal contributions of difference specific heat in case of these phases of pure elements.

7.3.3.2. Physical Models and Experimental Data

It is generally seen that the application of physical models in case of pure component make it possible to develop method of calculation of anharmonic characteristics (average temperature coefficient of expansion, Grunaisen parameters) for BCC- and FCC- phases pure elements from harmonic properties (equilibria volume, cohesion energy, and volume elastic modulus) with the help of equation state of Rose et al. The application of experimental data and physical models for explanation of changes of entropies elements between liquid, BCC-, and FCC- phase as function of temperature.

7.4. APPLICATION OF COMPUTATIONAL THERMODYNAMICS TO STEELMAKING AND PROCESSING

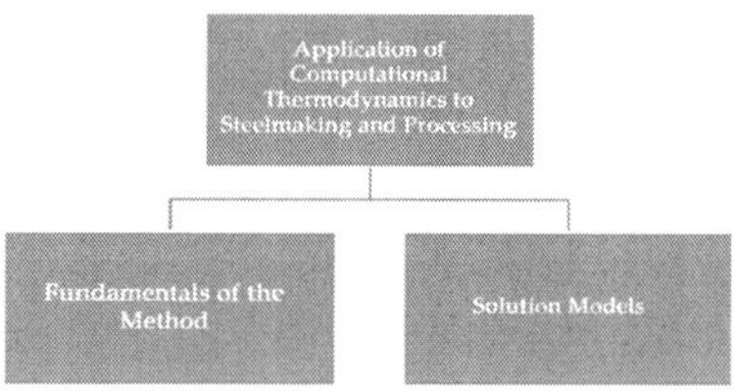

Figure 7.2. *Application of computational thermodynamics to steelmaking and processing.*

For more than a century, steel has been considered as the most important metallic material. Among materials that provide competition to the steel in some applications, only reinforced wood and concrete are produced in quantity higher than that of steel.

Despite the consistent claims to the doom of steel and the steel industry, the past few years have been observed as the noteworthy surge in the production of steel. Consequently, various other options that could act as a substitute for steels in many applications are being developed and offered every day. The primary reason that why steel has still maintained its supreme position in this segment is the potential capability of the industry to consistently innovate both in terms of products and in processing, facing various difficulties with steels that were not available previously or offering significant cost, environmental, and recycling advantages.

Albeit the beginning of the development of steelmaking and processing has been historically empirical, the present way to engage in production process to produce product thus do not follow this rule. While the "Physical Chemistry of Steel Making" committee of AIME was established in the year 1927, since the 1940's with the work of Lawrence Darken, John Chipman, and then of John Elliott, Carl Wagner and others, chemical thermodynamics became an essential element of the toolbox of steelmaking, reflected in classical works.

Furthermore, since the late 1960's with the work of Keith Brimacombe, Julian Szekely, and others, process modeling was introduced into the steel industry development too. Perhaps the major proof of this was the development of the AOD process.

With the increase in the challenges in the steel industry, these techniques have been widely adopted by various industries across different sectors. Furthermore, it became clear that there is need of a scientific, systematic, and efficient approach, if the challenges from other materials were to be addressed in time, within rational cost limitation.

The algorithms and mathematical approach developed in the last century for finding a solution to a physico-chemical problem were, evidently, restricted by the calculation tools available at the time. In times of logarithm tables and slide-rules, until the use of computers become more widespread, the solution algorithms had to compromise with the calculation limitations of the tools available at that time.

On the other hand, it was found that the physical metallurgy side of steel processing was primarily dependent on the graphical presentations of thermodynamic kinetics and equilibrium, mostly on the form of TTT or CCT curves and a phase diagram.

This, in some cases have the negative consequences as the diagrams or the solution algorithm that is used became so firmly fixed or resistant to change in a generation of metallurgists that it results in causing a confusion with the theory intended at supporting them.

Thus, at the time of using a completely different approach to the calculation of phase equilibrium (both in "physical metallurgy" and in steelmaking) with the help of computers viable, the approach was viewed with strong skepticism by most people in this field.

It is the result of the effort initiated by many pioneers such as Mats Hillert, Larry Kaufman, Himo Ansara and others to fight to establish that more comprehensive and more efficient solutions to phase equilibria problems could be possible with the help of the "computational thermodynamics" when the acronymic CALPHAD, "Computer calculation of phase Diagrams" was coined.

This effort was, after some period, become even more complicated by the parallel and somewhat independent development of the concepts and "classical" solution approaches, methods, and strategies to "phase diagram" problems and "thermodynamic problems" imperative for the practitioners in the two segments.

Furthermore, it is generally seen that sometimes the worthiness of understanding the equilibrium state of system in steelmaking as well as in processing has been ignored, in view of the believe that equilibrium was clearly not attained in most key circumstances.

However, as Mats Hillert has stated in the starting of CALPHAD developments, any sensible kinetic modeling should be constructed on the basis of thermodynamic driving forces and this can only be achieved appropriately, for industrial systems, by taking the help of computer calculations.

Whether one is in the position to attain equilibrium stage or reach near to it (for instance; in ladle steelmaking or in structures for high temperature application or) or to preclude it (e.g., in martensitic steels or in solubilized stainless steels), knowing what equilibrium stage is important in this field. Fortunately, in the past ten years, the use of CALPHAD by itself or as a part of kinetic modeling became widely popular and used by many in industry. The steel industry is actually the dominant player in this process.

Even within the current development of ICME, CALPHAD techniques are conceived as the prerequisite in the process design and computational materials.

Thus, five decades after the introduction of the CALPHAD method accompanied by the publication of the classical book by Nesor and Kaufman, there have been a notable evolution and industrial acceptance of the method.

7.4.1. Fundamentals of the Method of Thermodynamics and Computational Thermodynamics

The classical training of steel engineers in thermodynamics as well as in the physical-chemistry generally prioritize, on one hand, the thermodynamic theory and, while on the other, the methods adopted to find solution to various problems.

Unfortunately, it is common that the connection between the two facets are not highlighted, in special the importance of the connotation between the different essential thermodynamic properties. In a world of diagrams and tables, the engineer has gained experience in trying and finding the right data for the stated method. This is consistently inefficient and frustrating. If it is perceived that all thermodynamic functions are interlinked and undertakes that, for adequately high level of temperatures (above the Debye temperature, roughly, which is usually the instances for majority of steel problems) one can implement a simple polynomial description for Cp, proposed by Kubaschewski.

The relation between the variables relevant for steelmaking and processing is then quite evident, as shown in Eqs. (1)–(4).

$$C_P = m_3 + m_4 T + \frac{m_5}{T^2} + m_6 T^2 \quad (1)$$

$$H = \int C_P dT = m_1 + m_3 T + m_4 \frac{T^2}{2} - \frac{m_5}{T} + m_6 \frac{T^3}{3} \quad (2)$$

$$S = \int \frac{C_P}{T} dT = m_2 + m_3 \ln T + m_4 T - \frac{m_5}{(2T)^2} + m_6 \frac{T^3}{2} \quad (3)$$

$$G = H - TS = m_1 - m_2 T + m_3 T(1 - \ln T) - m_4 \frac{T^2}{2} - \frac{m_5}{2T} - m_6 \frac{T^3}{6} \text{ and}$$
$$\frac{\partial G}{\partial n_i} - G_i^0 = RT \ln a_i \quad (4)$$

Thus, in the initial development stage of the CALPHAD method, it was apparent that not all properties required to be stocked in a database, since all important properties could be reconstructed by just using one of them. While the integration constant of entropy (S) is linked with the Third Law of thermodynamics, the integration constant for enthalpy (H) is arbitrary. In the former time period, when most of the calculations were done manually or by slide rule, it is essential to indulge in right decision of a reference temperature to perform a process heat balance, without unnecessary performing of calculation.

This is not more the case, when most of the work are done by computers, and a single reference state could be completely adopted for the energy functions. The SER (Standard Element Reference) concept was standardized and implemented, since at time of introduction of Dinsdale's work and make it possible for steady data interchange between meaningful heat balances and databases, and enthalpy calculations.

Furthermore, most practitioners were also agreed about the of computational thermodynamics, that the Gibbs Free Energy (G) should be at the core in describing the databases. The practice of a polynomial description makes it possible for the computer derivation and integration. Evidently, it can be said that G is not an experimentally accessible property.

Thus, to create a database, it is important to use other experimental data. As it is well aware that all properties are interlinked with each other, and that equilibrium phase diagrams are the outcome of the reduction of G in the temperature, composition, pressure space, it is viable to comprehend that all this data can be used to get a reliable set of G for all relevant phases, to be included in a database, as presented in the "database box."

The process, which is considered as assessment- involves the accumulation and assessment of all available experimental data, the assortment of a meaningful solution model for G for the diverse phases and correcting the G functions by using an error minimizing function. Frequently, in this process, it is possible to discover the non-consistent data and the assessor must evaluate which data can be evaluated from the dataset.

The selection of a physically meaningful model for G is essential in a way to have a model that find the way for reliable extrapolations. Once there is availability of a database, reaching the equilibrium stage of the system becomes a mathematical challenge in minimizing the total Gibbs Free Energy G Total of the system, subjected to the limitation of mass conservation in the individual phases and in the system, as stated in the "equilibrium calculation box."

However, it may just appear like a straightforward problem, it is recommended to consider two aspects: (a) frequently G Total has local minima. The solution algorithm should be capable enough to being prevent from getting trapped in the local minima and reached the final equilibrium stage and (b) dissimilar from the "classical" approach of reducing G Total by looking the conditions for $\Delta G = 0$ for a given "reaction" (and thus, usually including only two phases- the forming precipitate and the steel, for instance), with which majority of engineers in steelmaking industry are familiar, the number of possible phases significantly increases (and accordingly the computational power required) with the increase in the number of components to a reasonable number (consider that, presently, the steelmaker industry frequently controls at least the contents of 6 essentials in a "normal" steel). With the maturity of computational thermodynamic, most of these problems are, to a major extent, appropriately addressed by several commercial computer codes.

Furthermore, for steels, significant effort was undertaken in forming very good databases. To get the maximum advantage from this technique, however, it is essential that the user recognize the new freedom from the classical algorithms.

While classical solutions of the steel engineer are generally restricted to problems in which the "conditions" are P, T, and composition, computational thermodynamics make it possible to use any thermodynamic variable as a condition, for as long as appropriate number of conditions is fixed to be able to solve the mathematical problem.

Furthermore, when one wish to have an idea about the liquidus temperature of a slag, it does not mean that one is in need of phase diagram of this system (and be restricted to only two-dimensional representations that will handle quaternaries with one component at a fixed level). Now, it become possible to actually regulate the liquidus temperature by calculation, in a high order system, without using a printed graph, which is the prerequisite for the steel engineers, and they are following this for decades.

So, the technique not only has to understood and learned, but it is also necessary to have a new frame of mind or perception in which one must question themselves about what information is actually required and not what type of diagram or table is usually looked up to find solution to this problem. This is one of the major challenges for those engineers that are recently introduced to this technique.

7.4.2. Solution Models

In the classical approach to thermochemical problems in steel making and processing one normally rely on a very limited number of solution models for which simple mathematical formulations are available and hence analytical solutions are possible.

Thus, Raoult, Henry (and Sievert) laws are usually applied and, in more extreme cases, the use of regular and sub-regular solution models is attempted (21), in special when one has to handle slags and similarly complex oxide-based solution. In general, however, the choice of the proper model is critical in obtaining a reliable description for G. In general, for a mixture or phase, G can be expressed as:

$$G^{\phi,T} = {}^{ref}G^{\phi,T} + {}^{conf}G_m^{\phi} + {}^{xs}G_m^{\phi} \tag{5}$$

In this equation the superscript "ref" (for reference) denotes the contribution of the un-mixed elements to the Free Energy, "conf" represents the configurational contribution (normally associated to the entropy change in mixing and a direct consequence of the physicochemical solution model selected) and "xs" denotes the excess Free Energy, the deviation between reality and the chosen model. The most frequent method to describe the excess free energy is an empirical polynomial following a suggestion by Redlich and Kister (22):

$${}^{xs}G^{\phi} = X_1X_2\sum_{v=0}^{n} L_v^T (X_1 - X_2)^v \tag{6}$$

where

$$L_v^T = A_v + B_vT \tag{7}$$

As a general rule, the smaller the number of terms in a phase description, it is more likely that there are higher chances that a description that which will be viewed as enough to allow for extrapolation into higher order systems and thermodynamic conditions when there is no data available.

7.5. COMPUTATIONAL THERMODYNAMICS, COMPUTATIONAL KINETICS, AND MATERIALS DESIGN

Generally speaking, CALPHAD is entirely the same method as computational thermodynamics and they are synonymous on most of the instance with little difference. The word "CALPHAD" recognize the importance of a technique in attaining a set of self-consistent model parameters from experimental means, empirical formulae, or first-principles calculations etc., and to establish multi-component and multi-phase databases.

For example, in the optimization module PARROT of Thermo-Calc software, it is very easy to choose a standardized phase diagram data, in addition to the thermodynamic data, decide weights for these data, and optimize the different data collected from variety of sources in one fitting procedure. Similarly, the model parameters of volume, atomic mobility, thermal conductivity, bulk moduli, etc., can be evaluated and databases can be determined within the framework of CALPHAD method. Inherited the thermodynamic models, the established databases and calculation methods from the CALPHAD method, computational thermodynamics stresses on engaging in the phase equilibria calculations and thermodynamics and applying these results in practical processes. Furthermore, computational thermodynamics can not only be considered as one segment of the multi-scale simulations but can also provide authentic data for simulations in other scales.

Moreover, in association with calculation of phase equilibrium from computational thermodynamics, computational kinetics can simulate microstructure evolution processes and phase transformations through using interfacial energy, diffusion coefficients, lattice constants, elastic moduli, etc., from physical property and mobility databases determined by the use of CALPHAD technique.

Simulation of kinetic processes is a complicated subject. Computational kinetics is very closely associated with the computational thermodynamics. It applies the Gibbs free energy and its first derivative in reference to the composition (for instance, chemical potential), the second derivative (for instance, the thermodynamic factor, also consider as the chemical driving force), and phase equilibrium data in order to simulate non-diffusional and diffusional phase transformations.

The bottleneck to evaluate and assess properties and performance of materials is the kinetic simulation of phase transformations and microstructure evolutions for multi-component and multi-phase materials systems. Some of the key methods in kinetic simulation for metallic alloys are discussed below.

7.5.1. Diffusion-Controlled Transformation Simulation

The commercial software DICTRA is considered as the most efficient tool in handling diffusion simulations for multicomponent and multi-phase oxide and alloys systems. DICTRA simulation is built within the context of Onsager's extension to multicomponent systems and on Fick's first and second law of diffusion. By inputting mobility/diffusivity data, and thermodynamic, employing numerical methods to find solution to partial differential equations for diffusion processes in single phases, DICTRA can simulate variations of flux, alloy concentration, and chemical potential with period in the diffusion zone.

It is important to note that the simulation of varying boundary within two adjacent phases is an iterative process. In each of the iterative step, DICTRA assumes that determining the local equilibrium is dependent at the sharp interface (i.e., varying boundary) and uses the concentration of calculated equilibrium as boundary conditions.

Then, the dissemination in every single-phase region is simulated and finally it become possible to determine the migration rate of the interface by the equation of flux-balance, which is basically the mass conservation law. For each of the component, it is not difficult to attain the one migration rate. Only when there is selection of proper interfacial composition, all rates for each and every component is identical, and the repetition terminates.

DICTRA simulation can be applicable only in case for planar, spherical, and cylindrical geometries, which are fundamentally one-dimensional diffusion process. It is a challenging task to handle more complex morphologies. There are many practical applications as follows: solidification process, homogenizing heat treatment; carburization and decarburization, segregation analysis; transformations between ferrite and austenite; nitridation and de-nitridation; sintering of cemented carbide; carbide growth and dissolution in steel; growth of pearlite in steel, etc.

7.5.2. Phase-field simulation

Unlike the above discussed method with the assumption of sharp interface, phase-field theory is based on the assumption that phase interface is diffuse. It introduces field variables, such as order parameter, composition, magnetic domains, and phase distribution, etc., to define the consistent alteration in the microstructure and to simulate the evolution of multi-phase microstructure by numerical solutions to evolution equations of conserved field variables (for instance composition) and non-conserved field variables (such as the long-range ordering, and crystal orientation, etc.).

This method does not require to consider the assumption of morphology, and microstructure, to trace the position of the interface. It is basically dependent on the space distribution of the field variables which progress or update with the minimizing process of the total Gibbs free energy.

This method is capable enough to adopted to simulate various diffusional and non-diffusional phase transformations, such as precipitation, solidification, coarsening, grain growth, dissolution, ferroelectric transition, martensitic transformation, ferromagnetic transition, crack propagation, dislocation dynamics, and so on.

7.5.3. Empirical Formulae

Presently, it is still not easy to predict microstructure evolutions and simulate phase transformations under several intricate processing and serving conditions solely by means of precipitation, nucleation, and kinetic theories. Thus, it can be said that there is need to search for semi-empirical or empirical methods from experimental data. Mehl–Johnson–Avrami model is one of the most widely used methods.

7.6. INTEGRATED COMPUTATIONAL MATERIALS ENGINEERING (ICME) AND MGI

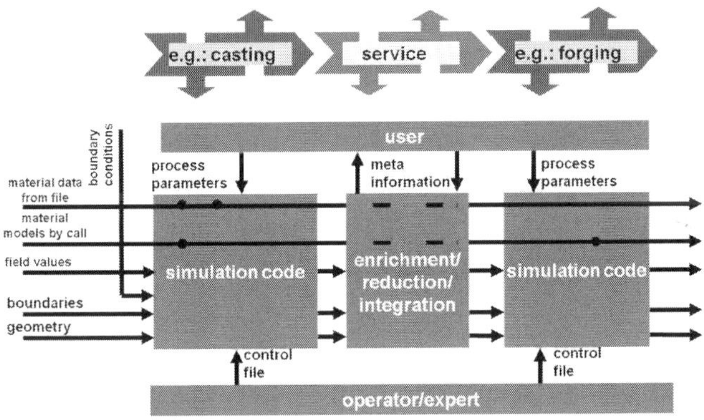

Figure 7.3. *Integrated computer material engineering concept.*

Source: Image by Wikimedia Commons.

ICME is a new concept that emerged recently aimed at integrating simulation of performance of materials, analysis of performance of engineering products, and integration of manufacturing processes into industrial system. This type of industrial manufacturing process generally chooses a unique procedure for different products, and the procedures play a crucial role on the microstructure of materials and on the overall performance.

The performance of materials is now not only dependent on the inputting constant involved in the designing of product, because the performance of the material changes in context with the determination of manufacturing process, thereby affecting the overall design of the product. In addition, the preliminary design of products could also transform with the improvement in the performance of materials. One of the key benefits of this integrated design is that the design of industrial products as well as performance of materials can achieve an optimal combination.

In June 2011, US President Barack Obama officially announced the MGI. There are a variety of understandings of "materials genome." Different understandings do not hinder the scientific research because the materials research is now inevitably developed into a new stage, i.e., combination of high-throughput experiments, high-throughput computation, and databases.

The success of the MGI relies on the systematically search for establishing the quantitative relationship among various materials composition/process-microstructure-property, in order to make sure that the rational design by means of computer simulation can trigger the rate at which material develops and deploy, thus reduce the cost and increase probability of success. Its overall objective and research ideas are similar to those of ICME.

It is important to note that there are several forms of materials and perhaps there does not persist any real materials genome that resembles biological genome. However, the central objective is to establish the relationship among composition/process-microstructure- property of materials.

This quantitative relationship exists through the entire life of materials from discovery, designing, producing, delivering, and recycling. This is the "materials genome" that one can understand and can be understood in different scales.

Computational thermodynamics at looking for the genome at both micro- and mesoscopic levels and trying to define the performance and properties of materials in a form of Gibbs free energy.

Successful engineering application of computational thermodynamics primarily rely on the authenticity of the dedicated databases, which in turn rely on a variety of correct experimental data, as well as anticipation from theoretical tools, such as calculations, computational thermodynamics, first-principles, etc.

Thus, it can be said that perfect combination of calculation-experiment-database depends on the computational thermodynamics, in reference with the MGI.

Computational thermodynamics method has been widely popular and part of every industry in the world. But combining them in order to form an integrated materials microstructure as well as for prediction of property system is still a new concept. The evolution of materials microstructure in practical production processes is still a very complicated task.

It is also seen that the temperature affects the overall phase transformation processes that are often conducive to the change of phase fractions and the discharge of latent heat which in turn have an effect on temperature.

Furthermore, the distribution of temperature in the industrial manufacturing processes is often non-uniform, so it can be said that the phase transformations in several parts of the products are different, which not only cause variations of the temperature field but also cause unequal distribution of stress induced by volume changes. It requires the integration of computational thermodynamics and finite element simulation through programming interface for real- time simulation and evaluation to attain reliable predictions.

In the near future, the integrated multi-scale calculations, along with the computational thermodynamics driven by dedicated databases, are proficient enough to stimulate the whole processes of production of materials and anticipating the properties of materials and microstructure, which make it possible precisely control production of materials during the planning and designing stage.

7.7. CONCLUSION

In the end, it is concluded that the computational that play the role of designing of metal-based phase diagrams, which dates back as early as previous century, has evolved in more recent years to the CALPHAD.

Computational thermodynamics is also conceived as an important element of materials informatics and is a cornerstone of the concepts behind the materials genome project.

While crystallographic databases are primarily used in context of a reference source, TD represent one of the earliest examples of informatics, as these databases were associated with the thermochemical computations to design or create phase stability in binary as well as in ternary alloys.

It is worth noticing that most of the computational thermodynamics concepts and software are designed by SGTE Group, a consortium which is conducive to the development of TD; the open elements database can be available easily as stated by the paper of Dinsdale. This so-called "unary" system resulted in acting as a common criterion for the construction of binary and multiple systems and is used by both open and commercial software in this field.

However, as described in recent CALPHAD papers and meetings, such a Dinsdale/SGTE database will require necessary actions in order to correct it over time despite the utility in keeping a common base. In this case, most published assessments will likely have to be revised, similarly, to rebuilding a house due to a severely broken foundation. This concept is also known by the name "inverted pyramid. Merely extending the current approach (restricted to the temperatures above room temperature level) is a complicated task. PyCalpahd, a Python library, was primarily created in order to facilitate simple computational thermodynamics calculation using open source code.

In complex systems, computational methods such as CALPHAD are used and applicable in model to thermodynamic properties for each phase and simulate multicomponent phase Behaviour. The application of CALPHAD at the time of high pressures in some important applications, which are not limited to one side of materials science such as the Fe-C system, authenticate experimental results by using computational thermodynamic calculations of phase relations in the Fe–C system at high pressures. Some other researchers even take into consideration viscosity and other physical parameters, which are beyond the domain of thermodynamics.

REFERENCES

1. academia.edu. (2013).*Application of Computational Thermodynamics to Steelmaking and Processing*. [online] Available at: https://www.academia.edu/27210197/Application_of_Computational_Thermodynamics_to_Steelmaking_and_Processing (accessed on 11 March 2020).
2. Anon, (2020).*Computational Thermodynamics: Calculation of Phase Diagrams using the CALPHAD Method*. [online] Available at: http://www.calphad.com/computational_thermodynamics.html (accessed on 11 March 2020).
3. Anon, (n.d.).*Thermodynamic Potentials*. [online] Available at: https://itp.uni-frankfurt.de/~gros/Vorlesungen/TD/5_Thermodynamic_potentials.pdf (accessed on 11 March 2020).
4. Application of Computational Thermodynamic and Solidification Kinetics to Cold Sprayable Powder Alloy Design, (2014). [ebook] Worcester Polytechnic Institute. Available at: https://web.wpi.edu/Pubs/ETD/Available/etd-011414–145401/unrestricted/Belsito.pdf (accessed on 11 March 2020).
5. Computational Thermodynamics and Informational Technologies in the Field of Computer Design of Materials, (2002). [ebook] Institute of Metallurgy and Materials Science after A.A. Baikov of RAS Moscow, Russian Federation. Available at: http://www.doiserbia.nb.rs/img/doi/1450–5339/2002/1450–53390204189u.pdf (accessed on 11 March 2020).
6. Computational thermodynamics: how to calculate phase diagrams without the fuss, (n.d.). [ebook] Available at: https://warwick.ac.uk/fac/sci/wcpm/seminars/warwick_lbp.pdf (accessed on 11 March 2020).
7. Computational thermodynamics: Recent developments and future potential and prospects, (2020). [ebook] Available at: https://onlinelibrary.wiley.com/doi/pdf/10.1002/pssb.201470107 (accessed on 11 March 2020).
8. Computational Thermodynamics: The Calphad Method, (2007). [ebook] Cambridge: Cambridge University Presss, New York. Available at: https://pdfs.semanticscholar.org/f18c/8e5970e47d3ea5079412c6e05bcdfc7f84da.pdf?_ga=2.253959114.379977704.1579155888–2084624878.1560189232 (accessed on 11 March 2020).
9. Folk.ntnu.no. (n.d.).*A Model for the Excess Gibbs Energy at High*

Pressures. [online] Available at: https://folk.ntnu.no/skoge/prost/proceedings/aiche-2004/pdffiles/papers/155f.pdf (accessed on 11 March 2020).

10. Introduction to Quantum Thermodynamics: History and Prospects, (2018). [ebook] Available at: https://arxiv.org/pdf/1801.08314.pdf (accessed on 11 March 2020).
11. Pubs.acs.org. (1979).*Functional group manipulation using organoselenium reagents*. [online] Available at: https://pubs.acs.org/doi/abs/10.1021/ar50133a004 (accessed on 11 March 2020).
12. Pubs.acs.org. (2019).*Accurate Computational Thermodynamics Using Anharmonic Density Functional Theory Calculations: The Case Study of B–H Species*. [online] Available at: https://pubs.acs.org/doi/full/10.1021/acsomega.9b00218 (accessed on 11 March 2020).
13. ResearchGate, (2002).*Applications of Computational Thermodynamics*. [online] Available at: https://www.researchgate.net/publication/273967017_Applications_of_Computational_Thermodynamics (accessed on 11 March 2020).
14. ScienceDirect, (2020).*Thermodynamic Assessment of the Ag–Au–Bi System*. [online] Available at: https://www.sciencedirect.com/science/article/abs/pii/S0364591606000745 (accessed on 11 March 2020).
15. SpringerLink, (2013).*Application of Computational Thermodynamics and Calphad in Magnesium Alloy Development*. [online] Available at: https://link.springer.com/chapter/10.1007/978–3-319–48194–4_1 (accessed on 11 March 2020).
16. The use of Computational Thermodynamics to predict properties of multicomponent materials for nuclear applications, (2012). [ebook] Available at: http://www.materials.cea.fr/MINOS2012/Session1/SUNDMANMINOS2012S1.pdf (accessed on 11 March 2020).

CHAPTER 8

Thermodynamic Modeling, Software, and Simulation

CONTENTS

The chapter of thermodynamic modeling, software, and simulation mentions the basic significance of the thermodynamic modeling, importance of the thermodynamic modeling, the various software's that are important to the thermodynamic modeling and simulation. The chapter briefly explains the thermodynamic modeling, selection of the Right Thermodynamic Models for Process Simulation, and substantiation of a thermodynamic modeling approaches, such as universal program ASTRA-4, and computer program TERRA, and Toolbox for the Modeling and Analysis of Thermodynamic Systems (T-MATS).

This chapter also provide highlights on the calculations of the phase equilibrium such as ideal model, equations of state model, activity coefficient models, along with the other various models. This chapter of the thermodynamic modeling, software, and simulation explains the significance of the thermodynamic simulation which is consisting of calculation of the thermodynamics, temperature effect and the heat transfer, optimum design of the thermal fluid systems such as heat exchangers, and analysis of the thermal effects with respect to the mechanical properties, electrical properties, and fluidic properties. This chapter also explains the modeling and the simulation of the thermodynamic systems such as Simulation Software for Thermodynamic Models.

8.1. INTRODUCTION

The study of phase diagrams has long been a very significant tool with respect to the development of the science and technology. In the context of metallurgy, it plays a very important role in designing the new alloys, partly due to significant enhancements in the experimental techniques which have reached a high level of precision.

Nevertheless, with the increment in the demands of new high technological material, which are commonly complicated, it is important that the theoretical anticipations assist the work of the materials scientists.

One of the possibilities is to put on the ideologies and philosophies of the physical-chemistry in order to calculate the multiphase equilibria and multi-component. With respect to this, a methodology can be very useful in both planning and also reducing the number of expensive experiments, considerably.

After the revolutionary work done by the well-known researcher named Meijering, which is followed by the significant contribution of researcher named Kaufman, great improvements have been observed, since the "NBS Workshop on the phase diagram," in the application of calculated phase diagram, that is developed with the help of underlying thermochemical data.

The main objective behind this is to improve the models with respect to the solution phases, and an increment in the amount of the consistent and assessed data is now being generated. The evaluation of the phase diagrams has been applied successfully in a wide range of applications, which is consisting of crystal growth of semi-conductor materials with the help of the liquid phase epitaxy, anticipation of phase equilibria with respect to the super-alloys, chemical vapor deposition, light metal alloys or high-speed steels.

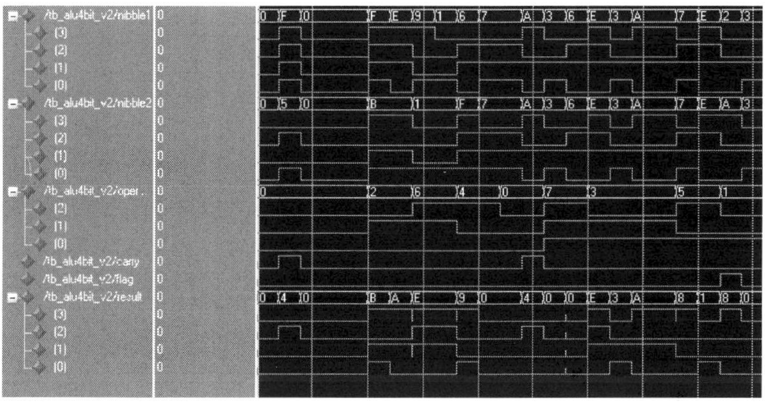

Figure 8.1. *An illustration of simulation of aluminum.*

Source: Image by Wikimedia Commons.

This has been possible because of the progress in the sophisticated application software. In the present interval of time, several numbers of application software are being used in various thermochemical computer data banks all across the world. In addition to this, a meaningful characteristic of phase diagram calculation is, the use of phase diagrams in the field of education in thermodynamics and material science in general.

Very simple computer programs have been put into use in order to depict graphically that how changing interaction parameters or lattice stabilities impacts the phase diagram. On the other hand, the more powerful application programs that are used in the simplest modes, can be used in order to illuminate the unreal simplicity of these kinds of concepts, for example, Gibbs Phase rule.

The assessment or evaluation of the phase diagrams needs the knowledge of the thermodynamics properties of the pure elements, the compounds and the solution phases, with respect to both stable as well as metastable physical states, and clear uniformity of these kinds of data between several types of phase systems are very necessary.

8.2. IMPORTANCE OF THERMODYNAMIC MODELING, SOFTWARE, AND SIMULATION

Process systems tools, for example, simulation, and optimization of the dynamic systems are extensively used in the process industries with respect to the development of the functional strategies and control for systems.

These kinds of tools are generally depending on the models of thermodynamics and in addition to that, several numbers of thermodynamic models have been created with respect to the mixtures and compounds.

Nevertheless, difficult models of thermodynamics are commonly computationally severe and are not available as open-source library with respect to the process simulation and optimization. In this topic, there are several numbers of application of novel open-source rigorous thermodynamic library are explained.

ThermoLib, which is mainly designed in the context of simulation and optimization of the vapor-liquid processes.

Mainly, the application of ThermoLib can be seen into the MATLAB and C.

ThermoLib uses the cubic equations of state in order to evaluate the various properties of the vapor and liquid phase thermodynamic. The distinctiveness of ThermoLib is that, it offers assistance and provide analytical first and second order derivatives.

These types of derivatives are required in order to have effective dynamic simulation and optimization.

The analytical derivatives provide assistance in order to enhance the computational performance by a factor between 12 and 35 as compared to the finite difference approximations. There are two basic examples are described with respect to the ThermoLib routines in their executions. First of all is simulation of a vapor-compression cycle, and second is the optimal control of an isoenergetic-isochoric flash separation process.

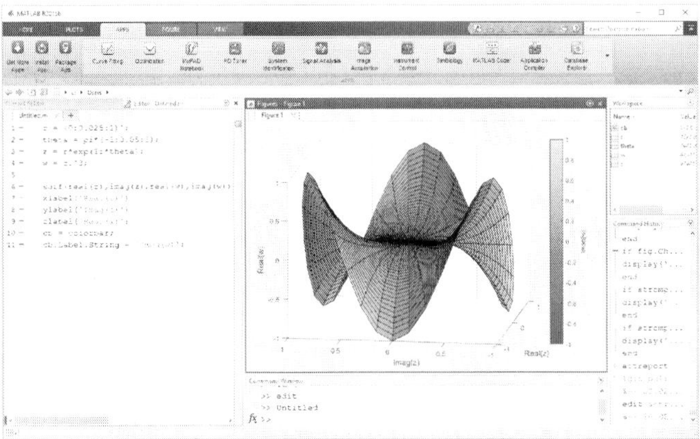

Figure 8.2. *An illustration of MATLAB software.*

Source: Image by Wikimedia Commons.

In the recent interval of times, the developments in the computer-aided modeling as well as simulation of complicated materials systems is consisting of three major components.

First of all, those basic tools of thermodynamics, state equations, must be revisited. Second is the new physical aspects and parameters of the materials modeling that require the increment in data precision obtainable with the help of recent physical as well as analytical instrumentation. And the last, but not the least is, requirement to develop highly specialized databases as tools in order to perform quality assessment, correlation, and prediction as well.

This chapter is consisting of concrete examples in metallurgy, composites, polymers, solutions, and gels – all that are relevant as well as necessary with respect to the materials scientists and engineers. It evaluates the behavior of the new materials and their complicated synthesis processes.

The increment in the significance of the CAD (computer-aided drafting) methodology in the context of Material Sciences is clearly identified. CAD (computer-aided drafting) contributions are viewed dynamically, for example, with respect to the materials life cycle.

8.3. THERMODYNAMIC MODELING

Thermodynamic modeling (TM) is a method of research that is being used all across the globe. TM is extensively used in the formation of ceramic and composite material, new metal and find out the new efficient, resource, and energy saving up the methods of their synthesis.

For instance, in order to evaluate the binary and multi-component diagrams of phase balance, which are a basic requirement for running out of new alloys and make one of the most significant and important bases in the context of material science, method CALPHAD (Calculation of Phase Diagrams) is used.

Modeling of an equilibrium status in multicomponent multiphase, the systems which are chemically reactive are being utilized to process out and optimization of several numbers of processes of synthesis of inorganic substances in materials technology and several numbers of various fields of chemical technology. In this way, the TM spend in several numbers of various modes, for instance, adiabatic, isobar-isothermal, etc. also use several types of computer methods and specialized programs.

TMpermits to explain the equilibrium composition of the interaction products (which are gaseous and condensed) in the multicomponent heterogeneous systems; temperature of chemically reacting system in adiabatic conditions (that is without taking into the account heat of loss in environment) in order to evaluate the impacts of several various factors on an exit of a product and to develop a basis for description of the probable mechanism of physical as well as chemical interaction, phase, and the structure formation in challenging system at synthesis and operation of new material.

With respect to the synthesis of the new materials, TM it has been successfully applied with the help of working out optimization of the reaction processes of multi-component powder filing the environment with the help of an aluminothermy approach for the subsequent formation of wear proof carbides coatings on carbonaceous steels with the help of thermochemical treatment.

In the present interval of time, the application of the TM is the important stage at formation of composite materials and the processes of their reception which is based on the phenomenon of self-extending high-temperature synthesis (SHS).

The main objective of the work was interaction consideration between the components of the heterogeneous powder compositions with respect to the titan and chrome oxides with the carbon on a surface of the alloyed steel at the atmospheric pressure, the formation of the equilibrium phase composition and anticipating the thermal stability of the titan and chrome carbides in an extensive interval of the temperatures which top value resembles to the melting point of a metal.

Modeling of the possible phase and chemical transformations at the process of self-extending high-temperature synthesis (SHS) and electron beam melting of the self-extending high-temperature synthesis (SHS) products on an alloyed steel surface at the formation of a composite membrane of the titan and chrome carbides of the thickness of near about 100 microns on air.

8.3.1. Selecting the Right Thermodynamic Models for Process Simulation

Software that permits the complicated processes to be simulated continues in order to develop for a wide range with respect to the thermodynamic conditions. Selection of the most suitable and useful models plays a very important role with respect to adequately representing real-life conditions.

One of the first and most important decisions to be made when modeling a chemical process system with the help of a computer simulation package is the selection of a suitable and useful method with respect to the simulating thermodynamic properties.

Thermodynamic-property methods permits the mathematical calculation of pressure, temperature, composition, and particular volume for multi-component as well as the multi-phase chemical systems. These calculation approaches are also used as a basis in order to determine the state functions of a system, for example enthalpy and entropy and in this way, Gibbs free energy.

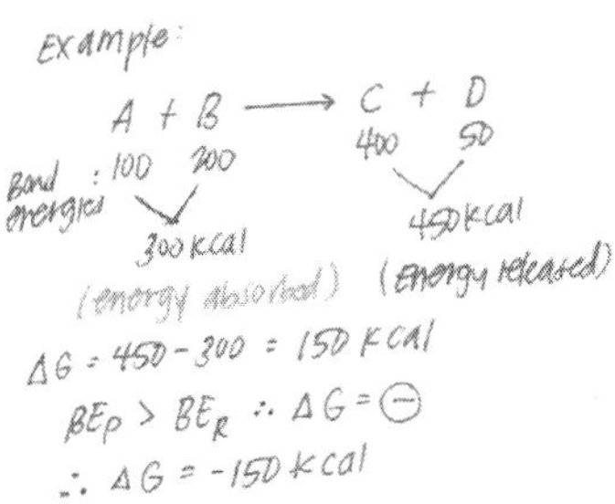

Figure 8.3. *An example of Gibbs free energy.*

Source: Image by Wikimedia Commons.

The topic of this chapter provides an overview of the primary thermodynamic-property methods, and explains the basis that underlies each one, the strengths and weaknesses of each, and applicability of each to several numbers of chemical systems, in order to provide assistance in the context of selection decision.

8.4. SUBSTANTIATION OF A THERMODYNAMIC MODELING TECHNIQUE

The adiabatic temperature Tad of interaction is a significant aspect in the self-extending high-temperature synthesis (SHS) processes. Last years for TM utilizes the computer programs with the thermodynamic databases (TD) which are based on the search of a minimum of Gibbs energy of multiphase multicomponent system.

In this way, TM compounds, which can be formed of the elements, which are a part of initial substances, and with respect to the formation of their quantitative parity responding balance under specific type of the conditions.

With respect to the self-extending high-temperature synthesis (SHS) processes in the multicomponent systems at the Institute of Structural Macro kinetics, Russian Academy of Sciences and issues of the materials technology of the Russian Academy of Sciences (Russian acronym ISMAN) established a specialized program named ISMAN-TERMO, which permits to spend TM only in the adiabatic mode.

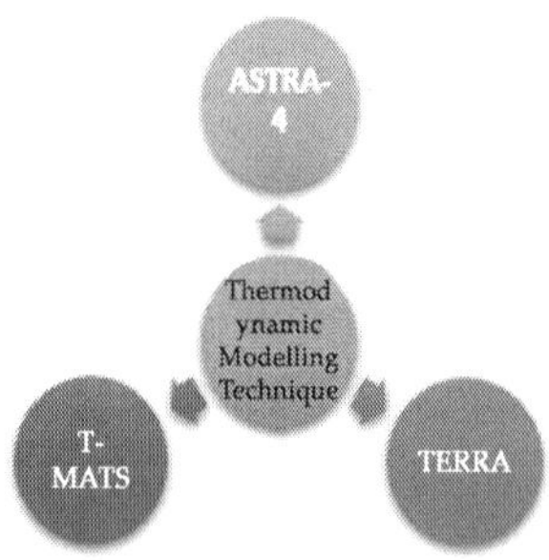

Figure 8.4. *Substantiation of a thermodynamic modeling technique.*

8.4.1. Universal Program ASTRA-4

Bauman state technical university has developed a universal program named ASTRA-4 that carries out search of the equilibrium composition with respect to a condition of a local maximum of entropy of system S in the existing constraints (a circumstance of weight continuation of each and every element and full internal energy of system) in order to set initial composition and a thermodynamic aspects and parameters.

TM of programs that are related to the ASTRA-4, these programs are based on the reference books on the thermodynamics of individual substances and is consisting of the big number of inorganic compounds.

The application technique of program ASTRA-4 with respect to the TM synthesis of materials in an adiabatic mode has been created in the earlier times with the reference to the conventional self-extending high-temperature synthesis processes and in order to synthesis of powder compositions with respect to the treatment of steels thermochemically.

It contains on carrying out of a calculation sequence, taking into the consideration about all possible binary diagrams and introduction to the TM of new compounds (absent in the standard configuration).

The configuration of the products received at system cooling after the process of synthesis from temperature Tad to room, can variate from the equilibrium composition at Tad.

In this way, additionally, the application of TM in the context of isobar-isothermal settings at the temperature below the points of phase transformations in the multicomponent system. At every level of the calculations are carried out at some point of time, that is, it is iterative, taking into the consideration that the temperature areas of existence of those or other phases products.

8.4.2. Computer Program TERRA

In the present interval of time, work for TM program TERRA that represents the new version of the program ASTRA-4 is used.

Program TERRA is proposed for work in the Windows System and possesses new chances with respect to the situation with TM, more expedient interface and the extended opportunities in order to present the calculations.

The software of complicated Terra with respect to the calculation of chemical reactions starts with several numbers of circumstances of balance: accomplishments of a minimum of Gibbs energy and an entropy maximum. In the first case, the optimizing problems presented in general, a type of condition ΔG_o reaction→min; ∑ N,n i,j Ci,jXj≤bi, (1).

Where 'bi' is the certain quantity of reaching the chemical element Bi; Xi is the stoichiometric factor of j compound, Ci, j is the quantity of the atoms Bi in compound j, N is the quantity of the chemical elements that take participation in the reaction; n is the quantity of the reaction products.

In the second case, the calculation is worked out, previous from the conditions of entropy maximum $S \rightarrow S_{max}$, a dependability of full internal energy, presentation of the continuation law of weight of all chemical elements, compliance of a gas phase to the composition equation for a mix of ideal gases.

A problematic formulation of TM demands to employ two types of conditions of balance of studied system with environment. These kinds of conditions can be or arithmetical values of thermodynamic aspect of equilibrium, or functional equalities between parameters of this status.

With respect to work, temperature, and pressure were fluctuate as parameters of the system equilibrium. For the explanation of the systems as material object, maintaining the chemical elements forming it was necessary the nobility only. Internal and interphase communications are explained with the help of modeling thermodynamic parities for which the short circuit properties only the individual substances such as equilibrium components are used.

The ease of the statement, problem that related to the program complex TERRA permits to use a thermodynamic method for studying of the big number of the diversified the high-temperature conditions and processes. According to the latest studies that used interface of a program complex:

Terra

The restraining number of chemical elements from which the inspected system can consist, is equals to fifty. The number of condensed phases is well thought out through the course of only one calculation, which is limited to two hundred.

The quantity of the components of the gas phase are formed in balance (which is the number of the individual substances), and it can be reach up to eight hundred.

With respect to performing the calculation part for heterogeneous systems probably use, both of the models of one-component not mixing up phases, and the models of the condensed solutions.

Triangle

Calculation can be done with two methods. First is the isothermal conditions when system equilibrium is set with the help of temperature (T, K) and the pressure (p, MPa). For the conditions of the adiabatic equilibrium transformation (burning) set in each and every point of the calculation with the help of pressure (p, MPa) and the enthalpy of reaction of initial substances (I, kJ/kg).

Rectangle

Areas with a similar set of the one-component reduced phases as an outcome of repetition of the calculations in the fixed range of temperature for all possible parities of the initial substances that come into the field of light. Thermal Behavior of the carbides, oxides, and nitrides that are being investigated in the temperature range from 273K (0.15 Degree Celsius) to 2673K (2400 Degree Celsius) at atmospheric pressure (105a).

Research of the stoichiometric compositions, in which the carbides of the materials that are transitive in nature are formed, carried out in the similar conditions. Possible interactions with the participation of Titanium (Ti), Carbon (C), Chromium (Cr) and the alloyed steel ($Fe_{70.88}C_{0.12}Cr_{18}Ni_{10}Ti_{1}$) in the atmosphere of the inert gas (Helium (He) and Argon (Ar)) and air ($N_{53.9}1O_{14.48}Ar_{0.32}$) were determined.

While performing the calculation, it is requited to well thought out about these phases, such as, nitrides TiN, CrN, Cr_2N, Fe_4N and Fe_3N_4, oxides Cr_2O_3, TiO_2, Ti_2O_3, Ti_4O_7, Ti_3O_5, TiO, FeO, Fe_3O_4, Fe_2O_3, NiO, and carbides TiC, Cr_3C_2, Cr_7C_3, $Cr_{23}C_6$ and Fe_3C.

The special consideration is given to the research of the phase equilibrium and in order to reveal the phase transformation sequence and the fields of crystallization of the co-existing phases in the system TiO_2-Cr_2O_3-C. phase balance with respect to the all concentration are with the help of one to five mol. Per cent were determined. Isothermal cuts, which are in the range of the temperatures from 773K (499 Degree Celsius) to 2573K (2299 Degree Celsius) and at pressure p=105Pa have been simulated.

In order to reveal the tendency of the process of crystallization of this or that phase, in a concentration triangle put isothermals crystalline fields at one pressure or on isothermal section that are in specified areas of phase taking into two account at different pressures. Developed materials are proposed for the reception of the composite heat resisting layers of carbides of the titan and the chrome, on a surface of the alloyed steels at non-vacuum electron-beam melting.

8.4.3. Toolbox for the Modeling and Analysis of Thermodynamic Systems (T-MATS)

T-MATS (Toolbox for the Modeling and Analysis of Thermodynamic Systems) is an open source software code. The T-MATS is a Simulink toolbox which is intended for the application in the modeling as well as simulation of the thermodynamic systems and their regulations.

T-MATS is consisting of several numbers of thermodynamic and regulatory components that can be integrated with a variable input iterative solver and optimization algorithm in order to develop the complicated systems to fulfill the requirements of the needs of a developer.

Enhancement of this tool was completed on behalf of the NASA Aviation Safety Program's Vehicle Systems Safety Technologies (VSST) project, which is located at the Glenn Research Center in Cleveland, Ohio.

8.5. PHASE EQUILIBRIUM CALCULATIONS

One of the most important calculations that are performed in the process of simulation is calculations of phase equilibrium. The basic principle or ideology of several numbers of unit operations (such as flash tanks, distillation columns, extraction, etc.) are based on the multi-phase equilibrium. Phase equilibrium can be evaluated with the help of the fugacity (which is a measure of the tendency of a component to leave its phase).

Equilibrium can be accomplished, when the fugacity of the component is same in all phases. The fugacity of the component *i* in liquid phase can be given by

$$f_i^L = \phi_{i,L} x_i P$$

And in vapor phase, it can be written as:

$$f_i^V = \phi_{i,V} y_i P$$

Where ϕ_i is the coefficient of fugacity. At equilibrium, it can be written as:

$$f_i^L = f_i^V$$

Obviously, with respect to the pure components of the fugacity, they can be simply reduced to $\phi_i P$, since x_i and y_i are both 1. The question here is how to determine the value of fugacity and the answer for that is relies on the system. In general, there are four choices, that are explained in the below section:

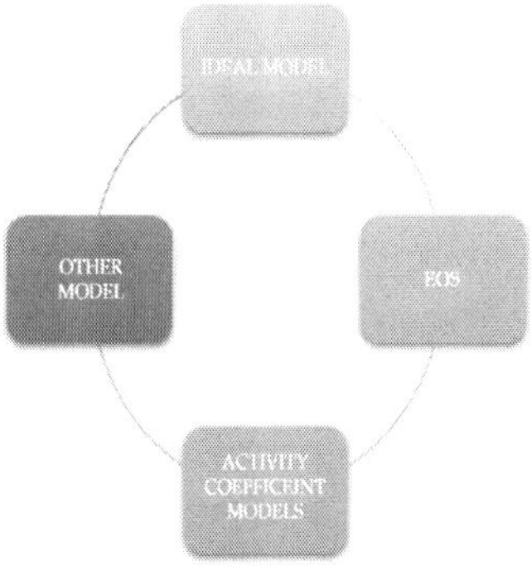

Figure 8.5. *Various models for phase equilibrium calculations.*

8.5.1. Ideal Model

An ideal system is consisting of an ideal gases and ideal liquids. The ideal fulfils the law of ideal gas, that is PV=n RT and has a fugacity of one. An ideal liquid has an activity coefficient (γ) of one. Ideal Behavior can be assumed for vacuum or low pressure or very high temperature operations, for gases, and when very small interactions (or interactions that cancel each other) in the context of liquids.

Interaction are insignificant when the molecules of the same size and the same character are combined together in the liquid phase. In Aspen Plus, ideal Behavior is modeled with the help of the Ideal property method. This method fixes the activity coefficient for the liquid phase to one, the equation of state (EOS) to the ideal gas law and estimates the molar volume of liquids with the help of the Rackett model.

Henry's law can also be used with the ideal model with the help of allocating the relevant components as Henry's component. According to the general thumb rule, when the system is consisting of the material such as organic acids, esters, water, amines, ketones, ethers, and aldehydes, then the individual is dealing with the polar molecules and there is a very good opportunity that the system deviates from the ideality considerably. For instance, the mixture of water and alcohol.

8.5.2. Equations of State Models

An equation of state (EOS) is a PVT relation that is applied to anticipate the properties of thermodynamics. Some of the equations of the state that are explained in the course of thermodynamics, for example, the cubic and the virial EOS's. With respect to the Aspen Plus, there are various numbers of equations of state that are used for various numbers of different application.

For instance, there the Peng-Robinson EOS (and its variations) and the Soave-Redlich-Kwong EOS (and its variations), which belongs to the cubic EOS. Other kind of the EOS is consisting of those derived from the statistical thermodynamics for example, the Sanchez-Lacombe and SAFT (statistical associating fluid theory). Another kind of the EOS models is the steam tables (that are provided as a base model in Aspen Plus).

8.5.3. Activity Coefficient Models

With respect to the non-ideal liquid solutions, the fugacity of the components in the solution deviated from that of the pure component. The ratio of the fugacity in the solution to that of the pure component can be explained as the activity. The activity that can be calculated from the activity of the coefficient (γ) as follows: In general, the activity coefficient is greater as compared to one.

What this means is that the fugacity of a component in mixture is higher as compare to that of pure component. In this way, the same liquid will have higher tendency, in order to vaporize when it is present in form of mixture as compared to its pure state. This can be attributed to the elevated repulsion between the molecules with the mixtures.

With respect to the few cases, the activity coefficient will be less than one. It shows the increment in the attraction between the molecules and less tendency to vaporize. With respect to the Aspen Plus, there are various numbers of models of activity coefficient. From which the most of them are commonly used is the NRTL (non-random two-liquid), which can be applied to the polar mixtures.

Other models are consisting of the Van Laar, UNIQUAC (universal quasi chemical), Electrolyte NRTL, Wilson, Flory Huggins, UNIFAC, and Scatchard Hildebrand models. In these kinds of models, the activity coefficient approach is applied in order to calculate the liquid properties. On the other hand, the vapor phase properties are determined with the help of the EOS.

8.5.4. Other Models

Still, there are several numbers of other property models available with the help of Aspen Plus. These models are categorized as "vapor Pressure and Liquid Fugacity Models" and have found applications in the specific processes.

From these models, there are API sour model (which is developed for the application of sour water treatment), Chao-Seader, and Grayson-Streed models (application to the heavy hydrocarbon systems), and kent-eisenberg model (for the amine sweetening units). Now, the question is how to choose a model. Generally speaking, several kinds of industries are tending to accumulate the experience on which model best fits its system.

For instance, from the experiences, models such as PR models and the SRK model for the oil and gas processing systems very well. In this way, these two models have been applied broadly for these kinds of systems. Other systems need specialized models because of the high non-ideality. For example, the amine sweetening unit in gas processing.

8.6. THERMODYNAMIC SIMULATION

Heat exchangers are applied in several numbers of energy as well as air-conditioning systems. Thermodynamic simulation is a convenient and a fast way to calculate the effectiveness and find out the optimum number, performance, dimensions, and position of the component with precision.

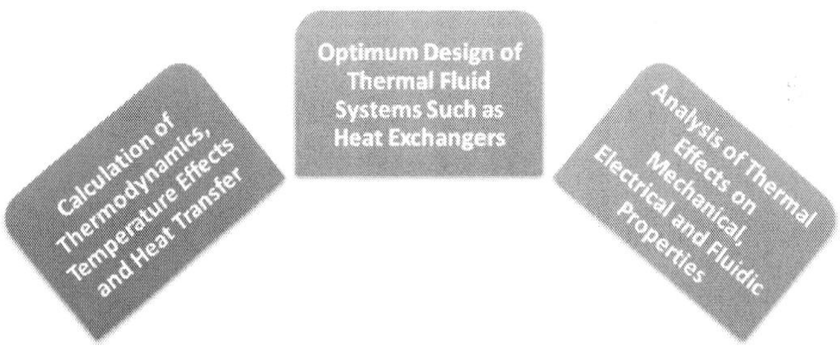

Figure 8.6. *Thermodynamic simulation.*

8.6.1. Calculation of Thermodynamics, Temperature Effects, and Heat Transfer

Processes such as energy and air-conditioning are impacted with the help of mass flow, coefficient of heat transfer, velocities, and the interdependencies between the temperature, state of fluid of matter, and pressure (for example, water, refrigerant, coolant, and air).

Heat exchangers are working on that basis, such as fins or plates heat exchangers, are mostly applied for example as counter-current or cross-flow heat exchangers in several numbers of devices and the system in a broad range of industries.

8.6.2. Optimum Design of Thermal Fluid Systems Such as Heat Exchangers

The most important elements of the thermal fluid systems, apart from the heat exchangers, pipes, values, pumps of fans, are condensers, expansion valves, capillary tubes and evaporators. With the help of the system simulation, an individual has the potential in order to find out the optimum number, position of the component, dimension of the component, and find out the most appropriate combination of the flow direction flow rates as well as pressure and the temperature ratios. This will provide assistance in order to increase the energy effectiveness of the products, save materials and decrease the required space for installation.

8.6.3. Analysis of Thermal Effects on Mechanical, Electrical, and Fluidic Properties

Elasticity, damping factor, density, magnetic properties, and viscosity, along with the electrical conductivity of the fluids and solids can alter or modify with the heat inputs or heat removal. This can be caused with the help of ambient conditions, but also with the help of process heat or heat loss within the machine. Also, an individual can run a reliable analysis of these kinds of effects with the help of a virtual simulation model of the physical system and the thermodynamic simulations.

8.7. MODELING AND SIMULATION OF THERMODYNAMIC SYSTEMS

Simulation software is very significant and effective tool, in order to find out the optimum number, dimension of the components, and the position of the components and the best possible configuration of pressure, mass flow and flow direction for the thermal fluid systems. An individual can also simulate the pressure drop of a gas at falling temperatures along with its energetic Behavior through the course of the transitions of the phase.

An individual can quickly create a thermodynamic model and evaluate the several numbers of configurations without any prototyping efforts. With the help of the thermal effects on the mechanical assemblies, electrical assemblies as well as fluidic assemblies, a person get a robust feedback on the Behavior of the machine under several numbers of ambient conditions, over the long operating periods or under the heat input from the mechanical as well as electrical work.

8.7.1. Simulation Software for Thermodynamic Models

The broad software, which is available with respect to simulating the systems, modeled with the help of standard bond graphs that does not adjust with most of the thermodynamic systems, and the reason behind this is that they are not modeled effectively with the help of these graphs. The convection bond graphs which adjust with these kinds of systems have not preciously been executed with the help of software of generic simulation.

The complications that are inherited in these kind of bond graphs, provides the assistance in the development of this kinds of software, quite difficult task. Heretofore differential equations for each and every model have been written on an ad hoc basis from the bond graphs.

Also, there are several downloadable software's that are free of cost, which provide assistance in order to deduce, organizes, and implements the differential equations for most convection and hybrid bond graphs. In this way, the simulating as an alternative to the Behaviour of lumped models that shows the hybrid thermodynamic systems, mechanical systems, electrical systems as well as hydraulic systems.

The imbedded computation of the thermodynamic properties of any of the thirty-six various single as well as multiphase substances is commonly precise and effective, which requires no interaction virtually. The user enters the binds graph largely with the help of two matrices, in this way, a graphical interface could be developed.

8.8. THERMODYNAMIC ANALYSIS AND CALCULATIONS OF (FE-CO) ALLOY BY MODELING AND SIMULATION USING THERMO—CALC SOFTWARE

Iron-cobalt alloy has the magnetic properties, with high Curie temperatures, the highest saturation magnetizations and high permeability etc. Researcher names Elmen discovered these alloys in the year of 1929.

Also (Fe-Co) alloy has been studied by the researchers named Ellis and Greiner. Normanton et al. in the year 2002 with the help of a researcher named Ohnuma. This alloy is used in the electric engine, which is consisting of a magnetic material.Thermo-Calc is powerful as well as flexible software and is used to perform several numbers of various kinds of calculations of thermodynamic as well the phase diagram. This software can be very significant in order to find out the solutions to difficult problems interaction of several numbers of elements and phases that shows highly non ideal behavior.The first version of thermo-calc was shown in the year 1981. In the year of 2002, modern version of thermo- calc was present. The first description on thermo-calc was done in the year of 1985.

8.8.1. Calphad Method

Full form of Calphad is calculation of phase diagram. The Calphad method is applied in order to find the thermodynamic properties of several numbers of various materials system. The calculation of phase diagram method was developed as a tool for phase equilibria of several numbers of various multicomponent systems and for treating thermodynamics. For the calculations, the modules reported are used.

CALPHAD (calculation of phase diagram) method has been used with respect to the material properties. In the present interval of time, the Calphad method has been utilized to a system which is having more properties that are based on the phases, as reported molar volume, elastic moduli as well as the database for processing these functions have been established.

For materials the Calphad method can be used to describing the composition, temperature of optical activity, thermo-electric and acoustic properties. All the calculations have been done by Calphad method.

8.8.2. Results

A thermodynamic calculation in the context of Fe-Co has been done with the help of a Thermo-Calc Software and Databases. The results of evaluated or assessed activities in (Fe-Co) at 1200˚K, 1225˚K and 1250˚K or 926.85˚C, 951.85˚C, and 976.85˚C, respectively. Gibbs energy for the investigated binary system, with respect to this topic, the reaction, equilibria, phase diagram and figures modules were used.

In liquid (Fe-Co) alloy their negative deviation can take place in activities. The activities show an increment proportionally with the increment in the temperature and deviation reduces in activity according to the Raoults law.

Activities show that there is strong interaction between the atoms of the alloy. Well thought out the calculated combined thermodynamic properties both the alloys show negative values for Gibbs energy which represents the stability of (Fe-Co) alloy according to Gibbs energy reduces with the increment in the temperature.

A thermodynamic property of Ferric base alloy system (Fe-Co) alloy has been found with the help of a Thermo-Calc software. It is shown that the thermo-calc software plays very vital role in the context of designing the structure and functional of Fe base alloy. The thermodynamic calculations in binary alloy (Fe-Co) at 1200 K, 1225 K and 1250 K were determined.

The binary system has negative values for Gibbs energy. The total Gibbs energy decreases with increasing temperature. At 1250 K, highest negative value of Gibbs energy in (Fe-Co) alloy system is observed which shows that the system is more stable, which increasing the level of stability, hardness, wear resistance, corrosion resistance and other doping characteristics of alloy. Activity values of the components are less than unity and show negative deviation from the Raoults law. Increase the temperature then activity increases and deviation in activity decreases, which indicates that there is strong interaction between the atoms of (Fe-Co). At 1250 K it has better accordance with Raoults law in Fe-Co binary system, which is important for engineering technology and metallurgy.

A Phase diagrams for the investigated binary system is calculated by the use of Thermo- Calc software and database. Obtained results presented one good base for the further thermodynamic analysis and developing in this group of innovative Fe base alloys. The obtained and shown results for the (Fe-Co) system provide complete thermodynamic optimization of this system.

8.9. CONCLUSION

This chapter concludes that Phase diagrams is used in order to find out or determine the binary system which can be calculated with the help of use of Thermo-Calc software and database. The obtained outcomes depicted one good base for the additional thermodynamic analysis and emerging in this group of innovative alloys which are made of Iron base. The obtained and shown outcomes for the (Fe-Co) system offer the complete thermodynamic optimization of this system. This chapter also conclude that there are several numbers of calculations models are present, that can be used in order to evaluate the phase equilibrium calculations.

REFERENCES

1. Andrew S. P., & Dr. Juan, C. A., (2020). Thermodynamics: Modeling and Simulation. [online] Mdpi.com. Available at: https://www.mdpi.com/journal/processes/special_issues/Thermodynamics_Modeling (accessed on 11 March 2020).
2. Ansara, (1990). *Thermodynamic Modeling of Solution Phases and Phase Diagram Calculations*. 62nd edn. [ebook] Great Britain: IUPAC, pp. 71–78. Available at: http://publications.iupac.org/pac/1990/pdf/6201x0071.pdf (accessed on 11 March 2020).
3. Atul, B. C., (2019). *Selecting the Right Thermodynamic Models for Process Simulation—Chemical Engineering | Page 1*. [online] Chemengonline.com. Available at: https://www.chemengonline.com/selecting-right-thermodynamic-models-process-simulation/ (accessed on 11 March 2020).
4. Brown, F., (2010). Simulation software for thermodynamic models, part 1. *Proceedings of the 2010 Spring Simulation Multiconference on—SpringSim '10*. [online] Available at: https://dl.acm.org/doi/10.1145/1878537.1878770 (accessed on 11 March 2020).
5. Cindy, S., (n.d.). *Thermodynamics*. [online] Simulationx.com. Available at: https://www.simulationx.com/system-simulation/physics/thermodynamics.html (accessed on 11 March 2020).
6. Claudius, P., (2014). *System Design, Modeling, and Simulation Using Ptolemy II*. 1st ed. [ebook] California, USA: Creative Commons Attribution. Available at: https://ptolemy.berkeley.edu/books/Systems/PtolemyII_DigitalV1_02.pdf (accessed on 11 March 2020).
7. Dr. Hussain, Y. A., (n.d.). *Thermodynamic Models & Physical Properties*. [ebook] Available at: http://www.just.edu.jo/~yahussain/files/Thermodynamic%20Models.pdf (accessed on 11 March 2020).
8. Kornilov, S., Rempe, N., Smirnyagina, N., Khaltanova, V., & Lapina, A., (2016). Thermodynamic Modeling of High temperature Synthesis of the Titan and Chrome Carbides on an Alloyed Steel for Electron-Beam Melting of Modifying Coatings. *MATEC Web of Conferences*, [online] *79*, p. 01033. Available at: https://www.matec-conferences.org/articles/matecconf/pdf/2016/42/matecconf_imet2016_01033.pdf (accessed on 11 March 2020).
9. Ritschel, T., Gaspar, J., & Jørgensen, J., (2017). A thermodynamic library for simulation and optimization of dynamic processes. This

project is funded partly by: 1) innovation fund Denmark in the CITIES project (1305–00027B) and in the OPTION project (63–2013–3), 2) the interreg project smart cities accelerator (10606 SCA), and 3) EUDP 64013–0558 in the IEA Annex for energy efficient process control. *IFAC-PapersOnLine,* [online] *50*(1), 3542–3547. Available at: https://backend.orbit.dtu.dk/ws/portalfiles/portal/139332781/1_s2.0_S2405896317314234_main.pdf (accessed on 11 March 2020).

10. Software.nasa.gov. (n.d.). *Software Details*. [online] Available at: https://software.nasa.gov/software/LEW-19165–1 (accessed on 11 March 2020).
11. Syed, M. S., Tariq, U., Muhammad, S. K., Nasib, U., & Bakhtar, U., (2019). *Edelweiss Publications*. [online] Edelweisspublications.com. Available at: http://edelweisspublications.com/articles/23/218/Thermodynamic-Analysis-and-Calculations-of-(Fe-Co)-Alloy-by-Modeling-and-Simulation-using-Thermo—Calc-Software (accessed on 11 March 2020).
12. Thermodynamic Modeling and Materials Data Engineering, (1998). *Data and Knowledge in a Changing World*. [online] Available at: https://www.springer.com/gp/book/9783642722097 (accessed on 11 March 2020).
13. Thoma, J., & Bouamama, B., (2000). Modeling and Simulation in Thermal and Chemical Engineering. [online] Available at: https://www.springer.com/gp/book/9783540663881 (accessed on 11 March 2020).

CHAPTER 9

Application of Thermodynamics

CONTENTS

In this chapter, the application of thermodynamics has been discussed in detail. The application of second law of thermodynamics is in use very much in the applications that human being uses in our day to day life. The application of thermodynamics in automobile engines has also been discussed.

In this chapter the application of thermodynamics in thermal machines has also been discussed particularly in heat pumps and refrigerators. The applications of thermodynamics in context of sustainability has also been discussed.

In this chapter the use of thermodynamic chemistry in food chemistry has also been discussed in detail.

9.1. INTRODUCTION

The basis of the second law of thermodynamics was laid by discoveries made by Sadi Carnot, a young French scientist known to be the father of thermodynamics. Prior to Carnot's findings it was believed that heat engine efficiency relied on the form of working fluid used in the engine. In his book, Reflections on Fire's Motive Power published in the year 1824, Carnot said the heat engine's output was independent of the type of fluid used in the engine.

According to him, heat engine efficiency depends only on two temperatures: the source temperature (hot body) from which the engine absorbs heat, and the sink temperature (atmosphere) from which the engine gives up the exhaust heat. It was a great discovery, but it was not good to people that the engine's output relied only on the source and sink's high and low temperatures, respectively.

Findings of Sadi Carnot have been overlooked, and many of his articles have gone unpublished. His book was luckily written. Future scientists such as Kelvin reviewed the observations of Carnot in more detail and made more progress. The definition of the second thermodynamics law applied to heat engines applies equally to the internal combustion engines used in our vehicles, bikes, ships, aircraft, etc.

In internal combustion engines heat is produced inside the engine by combustion of fuel. The combustion of fuel occurs as a result of spark generation (Spark Ignition or SI engines), as in the case of gasoline engines or as a result of fuel compression (Compression Ignition or CI engines) as in the case of diesel engines.

Some part of the heat generated inside the engine is used to conduct the piston inside the engine cylinder going to work. The piston is attached via connecting rod to the crankshaft. The piston's reciprocating motion is converted into the crankshaft's rotating motion, which is converted via gearbox to the rotating motion of the wheels.

The remaining part of the heat generated inside the engine is emitted as the exhaust gases or tailpipe emissions into the atmosphere. In this case the engine is called the source where heat is produced, whereas the atmosphere where heat is emitted is called the drain.

According to the second law of thermodynamics, when the source temperature is higher and the sink temperature is lower, then the engine performance is higher.

The second thermodynamics law applies to all engine cycles like Otto, Diesel, etc., irrespective of what sort of gas they are operating on and what type of fluid. It is the discovery of this law that has contributed to humanity's advancement towards the automobiles of the present day. There is basically no other law which applies as extensively as the second thermodynamic law.

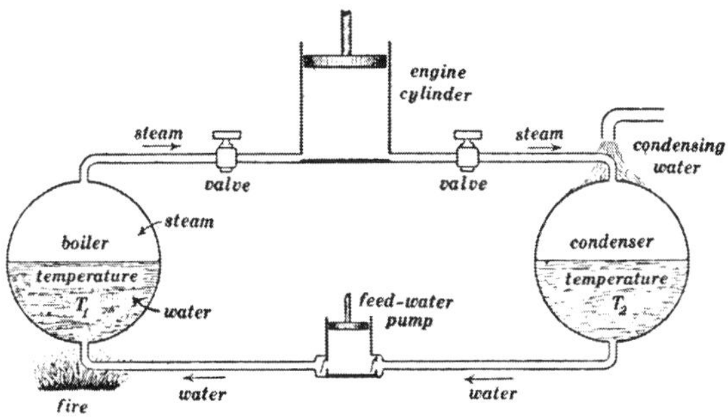

Figure 9.1. *Demonstration of second law of thermodynamics.*

Sources: Image by Wikimedia.

9.2. THERMAL MACHINES

A machine is a moving parts mechanical device that helps to perform any useful operation, typically from a basic wheel-and-shaft to a nuclear power plant. Machines can be just mechanical (pulley-and belt), hydraulic (water wheel, earliest engine), pneumatic (windmill), electric (electric motor, nowadays most common), chemical (fuel cell), or thermal. The name thermal machines can be limited to devices that convert heat to work in a heat engine, and to devices that pump low-temperature energy to high-temperature energy (using some exergy source, all operating in a cyclical manner.

The heat pumping system may be designed to produce cold that is refrigerator, or heat or heat pump, or both simultaneously that is heat recovery refrigerator or heat pump. It has been observed that engine thermodynamics is a broader subject than Heat-Engine thermodynamics; e.g., fuel cells are not heat engines but electrochemical engines, and internal combustion engines are really Thermal machines and heat engines thermochemical engines, their study requiring Chemical Thermodynamics; in the strict sense, only heat engines are steam engines and Stirling engines, and the like. Within Thermal Systems, there is a wider view of generating power.

9.2.1. Application in Heat Pumps and Refrigerators

Almost every home is equipped with a fridge. Many people don't realize they're sharing a heat pump with their homes too.

Heat pumps, air conditioners, and fridges use heat transfer from cold to heat. They are reverse heat generators. They say reverse, rather than reverse, because all heat engines, although they can be run backwards, cannot be completely reversed except for the Carnot engines. Heat transfer is from and into a cold reservoir. It includes work input, which is transformed into heat transfer as well.

Thus, the flow of heat to the hot tank is. The function of a heat pump is to transfer heat to a warm environment, such as a winter home. Air conditioners and refrigerators have a function to allow heat transfer from a cool environment, such as cooling a room or keeping food at lower temperatures than the weather.

A heat pump can actually be used for both heating and cooling a room. It is basically an air conditioner and a heating machine all in one. In this section we will concentrate on its heating mode. Heat pumps, air conditioners and refrigerators are heat engines running backwards. The one shown here is based on a (reversible) Carnot motor.

- Showing heat transfer from cold tank to warm tank with heat pump. instructions, and what they would be in a heat engine are opposite.
- Carnot cycle diagram similar to that in but reversed, following ADCBA direction.

The region inside the loop is negative, which means there is a net contribution to the job. There is heat transfer from a cold reservoir down path DC into the system, and heat transfer from the system to a hot reservoir along path BA.

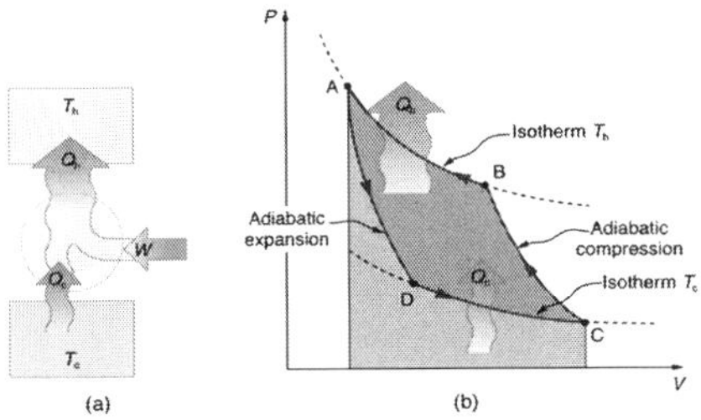

Figure 9.2. *Graph of a heat pump working.*

Source: Image by Opentextbc.

9.2.2. Heat Pumps

The great advantage of using a heat pump to keep your home warm instead of just burning fuel is that it is provided by a heat pump. Heat transfer is from outside air to indoor space, even at a temperature below zero. One only pays for, and then receive an extra heat transfer from the outside at no cost; in many situations, at least twice as much energy is added to the heated room as the heat pump is used to operate.

One must pay for all this when he burns fuel to keep it dry. The downside is that the input of work (required by the second law of thermodynamics) is sometimes more expensive than simply burning fuel, particularly when the work is done by electric power.

A work fluid such as non-CFC coolant is used. In the outdoor coils (the evaporator), heat transfer occurs from the cold outdoor air to the working fluid, transforming it into a flame.

There are four basic components of a single heat pump:

- condenser;
- expansion valve;
- evaporator; and
- compressor.

In the heating mode heat transfer occurs from the cooler outdoor air to the working fluid in the evaporator (3), rendering it a gas. The electrically driven compressor (4) increases the temperature and pressure of the gas and drives it inside the heated space into the condenser coils (1). Because the gas temperature is higher than the in-room temperature, heat transfer from the gas to the room occurs as the gas condenses into a liquid. The working fluid is then refrigerated as it flows back to the outdoor evaporator coils through an expansion valve (2).

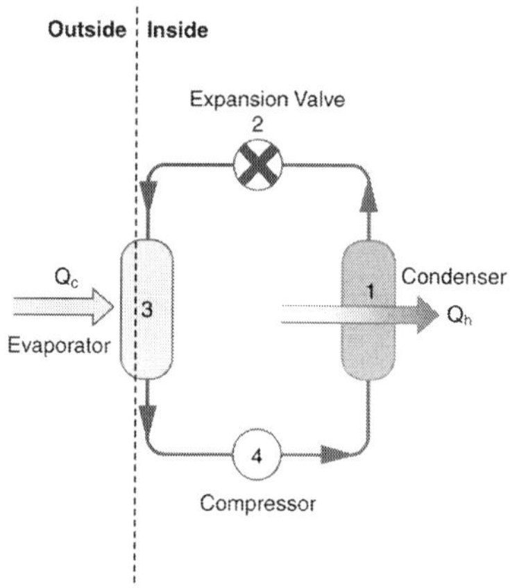

Figure 9.3. *Working of heat pump.*

Source: Image by Opentextbc.

The electrically driven compressor (work input) increases the gas temperature and pressure and drives it into the inside of the heated room condenser coils. Because the temperature of the gas inside the room is greater than the temperature, heat transfer happens to the atmosphere and the gas condenses to a liquid.

The liquid then flows back to the outdoor evaporator coils via a pressure-reducing valve, being cooled by expansion (In a cooling cycle, the evaporator and condenser coils switch positions and the fluid's flow path is reversed.). The efficiency of the heat pump is determined by how much heat transfer occurs in the warm space compared to how much work input is needed.

In the spirit of taking the ratio of what you get to what you pay, we consider the output coefficient) (of a heat pump to be because a heat engine's efficiency is, we see that, an important and interesting fact. Second, since any heat engine's output is less than 1, it means that it is always greater than 1—that is, a heat pump always has more heat transfer than the work put in it.

Second, it means heat pumps work best when variations in temperature are small. The efficiency of a perfect engine, or Carnot, is; thus, the lower the difference in temperature, the lower the output, and the greater the difference (because). In other words, heat pumps do not operate in very cold climates as well as in warm climes.

Friction and other irreversible processes decrease the output of the heat engine but do not support the operation of a heat pump–rather, they minimize the work input by converting some of it to heat transfer back into the cold reservoir before it reaches the heat pump.

When a real heat engine is running backwards, some of the intended work input is converted to heat before it reaches the heat engine, thus raising its output coefficient. In this figure, the portion of that goes into the heat pump is reflected, while the remainder is lost to the cold reservoir in the form of frictional heat.

If everything had went into the heat pump then it would have been bigger. The best heat pump uses adiabatic and isothermal methods, since ideally there would be no dissipative processes to minimize the transfer of heat to the hot tank.

Actual heat pumps in the previous example do not perform quite as well as the ideal one; their range values from around 2 to 4. That range means the heat transfer from the heat pumps is 2–4 times as high as the work put into them.

However, their economic feasibility is still limited, as it is usually supplied with electrical energy which costs more per joule than heat transfer by burning fuels such as natural gas (NG). In fact, the initial cost of a heat pump is greater than that of many furnaces, so a heat pump must last longer to recover its cost. Where winter temperatures are moderate, electricity is relatively cheap and other fuels are relatively expensive, heat pumps will most likely be economically superior.

As they can cool as well as heat up a space, they also have advantages where cooling is also needed in summer months. So, some of the best heat pump sites are in warm summer climate with cool winters. This shows a heat pump, in some countries called a reverse loop or split-system cooler.

Heat transfer occurs during hot weather from air inside the room to air outside, cooling the room. During cool weather, heat transfer takes place from outside air to inside air, warming up the room. Such switching is accomplished by inverting the working fluid's flow path.

9.2.3. The Motion Power Engines and Motors

They are used as stand-alone plants to deliver motion or electricity to other systems, or within vehicles to provide motive power (propulsion) and auxiliary energy. In 2000, there were approximately 109 billion vehicles around the world, more than 80% of which were passenger cars. Between 1900 and 2000 more than million motor vehicles were produced, and a similar amount is expected to be delivered between 2000 and 2020.

The steam engine is now only used in the largest power plants that is nuclear and coal-fired, and the most common heat engine today is the reciprocal internal combustion engine for automobiles, buses, ships, small thermal machines and heat engines aircraft, and stationary engines, with a third type, the gas turbine engine gaining ground for most aircraft, the faster types of ship, and combined power-and-heat stations.

The largest thermal power plants are vapor turbines, usually limited to 1000 MW per unit in nuclear power plants due to reactor heat transfer limitations that is fuel power plants are restricted to about 400 MW per unit due to limitations in combustion intensity.

Gas turbines can also achieve some 300 MW per cycle, and the largest reciprocating engines are up to 100 MW of marine diesel engines (the largest petrol engine is just 0.3 MW). Remember that its power is always to be interpreted as its output power when referring to a heat engine (i.e., its mechanical or electrical power; the thermal power input, usually referred to as the low heating value of the fuel used, is usually three times greater).

9.3. THERMAL ASPECTS OF HEAT ENGINES

The experts have assessed the internal workings of heat engines, but only from the aspect of thermodynamics, i.e., the type of thermodynamic processes take place in actual heat engines, but not considering other important aspects of heat engines, whether thermal or non-thermal.

Thermal problems in power generation equipment typically involve heat transfer problems, especially problems of cooling, that are often important for all types of engines, not only in high-temperature turbine blades for gas turbines, valves, and walls cooling in reciprocating engines, but also in more general heat dissipation problems due to friction, sealing problems and thermal expansion interference like losses, wear, noise, cracks, etc.

In addition, the study of heat engines is usually linked to the study of fuels and their actual combustion processes, the materials involved (including the required cooling), the mechanisms and their lubrication, the piping of the working fluid that is air intake, fuel injection and product exhaust, the ignition if any, the structure and its supports (including vibration insulation), their performance in terms of shaft speed and torque (e.g., for coupling to vehicle wheels or screw propellers), the electronic system (sensors and controls), testing, and diagnostic means, etc.

And each of the above aspects may require ever-increasing analytical attempts; for example, in a few milliseconds, a thin fluid stream in turbulent motion can cavitate and flash boil at the tip of the fuel injector, form a spray, break it down to a myriad of droplets that evaporate, ignite, burn, and produce unwanted emissions. It is known that knowledge of the basic principles of heat engines is not enough to develop a working machine, not even to make it run.

Though a more general presentation of heat engines is provided here, the proposed exercises are essentially limited to ideal cycles with few practical modifications (e.g., turbomachinery isentropic efficiencies, heat exchanger temperature shift, and so on

9.3.1. Carnot Cycle

With his masterpiece Réflexions sur la puissance motrice du feu, et sur les machines propres à developer cette puissance (Reflections on the motive power of fire, and on the machines capable of developing this power) in 1824, Nicolas Leonard Sadi Carnot was the first to propose a thermodynamic model of a heat engine, abstracting from the only available heat engine, the steam engine, to describe the fundamentals: the concept of a generic functioning fluid, conducting a generic cyclic process, interacting with generic heat reservoirs.

In that his only publication, Carnot concluded that all heat engines are limited by the operating temperatures in their energy conversion efficiency and that maximum efficiency is achieved when the operating fluid is assumed to follow four ideal processes:

- thermal machines and heat engines;
- an isotropic compression (1 to 2) to change temperature without heat transmission;

- the isothermal heat input (2 to 3), from the hot source at the temperature of the hot source;
- the isotropical expansion (3 to 4) to change temperature without transfer of heat; and
- rejection of isothermal heat to the cold source usually environment at temperature of the cold source.

Carnot concluded with a set of rational deductions, namely: any friction engine would have less efficiency than one without, across all engines exchanging heat at different temperatures, the one with the highest efficiency only exchanges heat at two extreme temperatures the hottest and coldest and all reversible engines operating at the same extreme temperature pairs.

The energy and energy efficiencies have been established and can be easily deduced by setting the total energy conservation, $\Delta_{Euniv} = Q_{hot} - W - Q_{cold} = 0$, and the total entropy balance, $\Delta S_{univ} = -Q_{hot}/T_{hot} + Q_{cold}/T_{cold} \geq 0$. The latter being zero in the ideal case of a Carnot cycle, that gives:

$$\eta_e \equiv \frac{W_{net}}{Q_{pos}} \rightarrow \eta_{e.Carnot} = 1 - \frac{T_1}{T_2}, \eta_x \equiv \frac{\eta_e}{\eta_{e.Carnot}} \rightarrow \eta_{x.Carnot} = 1$$

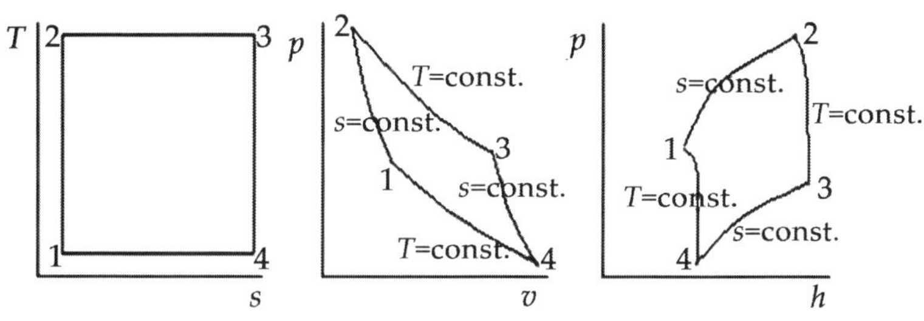

Carnot cycle and any other conceivable power cycle shall be executed in clockwise direction in both T-s and p-V thermodynamic diagrams, since the heat engine receives net heat, that in a reversible process is Qnet= ∫TdS (recall area interpretation of integrals), and delivers network, that in a reversible process is Wnet= ∫pdV; notice that in a cycle ∫dU=∫TdS−∫pdV=0.

The Carnot cycle is not practical, not only because of inevitable frictional losses (which could be minimized with adequate lubrication), but also because of the heat transfer with negligible temperature jump, which would make the heat transfer rate infinitesimal for a finite size engine with finite heat transfer with heat sources.

Heat transfer rates, however, can be greatly increased if the gas passes through a porous medium that acts as an intermediate heat source, as in the Stirling and Veuillemier cycles.

9.3.2. Gas Power Cycles

A working fluid cyclically evolves inside a heat engine by taking heat from a hot source and rejecting heat into the atmosphere, thereby generating some mechanical work in that manner that is generally shaft work. Thanks to its fast absorption of thermal energy by compression or expansion, gases are the best working substance while liquids have little compressibility.

These are called gas cycles where the working material remains in the gas phase all the time, and vapor cycles those where the gas in some portion of the cycle condenses to liquid and back again. While any cycle can in theory be used as a heat engine or as a cooler and heat pump by simply reversing the direction of the flow, there is a big difference in practice and the analysis is split between power cycles and refrigeration cycles.

Several gas cycles have been proposed and many are currently being used to design real heat engines: the Otto cycle takes into account the actual petrol engine, the Diesel cycle average the actual diesel engine, the mixed cycle is better than both, a hybrid of the last two, the Brayton cycle approximates the actual gas turbine engine very efficiently, and the Stirling cycle used in some exotic applications.

The study is limited to the so-called air normal model, that assumes air with temperature-independent properties as the working fluid, neglecting the effects of fuel addition the air to fuel mass ratio is close to 15:1 in Otto engines, nearly 30:1 in Diesel engines, and nearly 40:1 in Brayton engines), and an equivalent heat addition is taken into consideration.

9.3.3. Otto Cycle

The Otto cycle is the first attempt to design the operation of a spark-ignition engine, first built in 1876 by Nikolaus Otto and used in many cars, small planes and small power systems that is below to 200 kW down to miniature engines.

This is a dual internal combustion gas engine that is the master in the twentieth century; in the nineteenth century, it was the external combustion steam engine. Usually the Otto engine uses gasoline as fuel, though can also function on liquefied petroleum gas (LPG), natural gas (NG), alcohol, etc.

It is described in figure where the usual terms for engine-geometry characteristics (stroke, bore, displacement, and compression ratio) are used; other terms for engine-operation characteristics include shaft speed, mean pressure, strength, fuel consumption, torque, volumetric efficiency, energy efficiency etc. The intake has conventionally been a homogeneous combination of air and gasoline, but direct injection of fuel into the cylinders takes over.

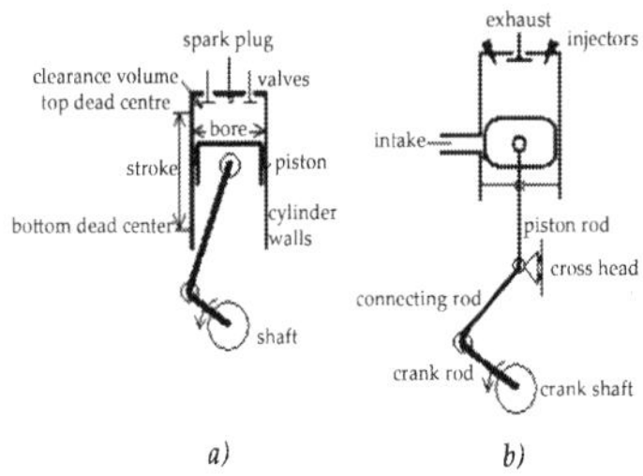

Figure 9.4. *Depiction of working of an engine on Otto Cycle.*

Source: Image by Webserver.

In the ideal air-standard Otto cycle, the working fluid is just air, which is assumed to follow four processes

- isentropic compression,
- constant-volume heat input from the hot source,
- isentropic expansion,
- and constant-volume heat rejection to the environment.

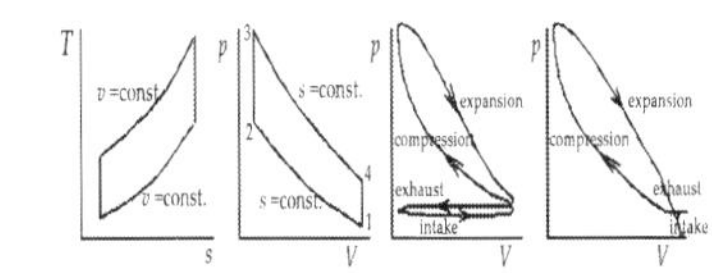

The ideal Otto cycle in the *T-s* and *p-V* diagram, and a practical *p-V* trace of four-stroke and two-stroke engines.

The key parameters of both ideal and real Otto cycles are:

- Size determined by volume of displacement (piston volume, V1−V2), generally less than 0.5 liters per cylinder, to avoid self-ignition.
- Speed crankshaft speed n more specifically, with a standard operating range n = 1000–7000 rpm (n = 20–120 Hz).

For four-stroke engines the maximum value can be in the range n_{max}=6000–8000 rpm. Two-stroke motorcycle engines run faster (n_{max}=13,000 rpm), the fastest (n_{max}=20,000 rpm) being the smallest engine that is used aircraft modeling engines (two-stroke, 1 cm^3, 200 W, by use of methanol or ether fuel with some 5 to 15 % lubrication and anti-corrosion oil.

Ratio of compression, r = V1/V2, with a usual range of r =8. 11 (up to 14 in direct-injection spark ignition engines), restricted by knock or self-ignition issue.

Mean effective pressure, p_{me}, described as the unit work divided by the displacement with a usual range of 0.2–1.5 MPa (the full load value may vary from p_{me}=1.2 MPa in two-stroke motorcycle engines to p_{me}=1.7 MPa in the largest turbocharged engines).

Maximum pressure could have a standard range of 4–10 MPa scale. The Reciprocating engine performance maps are normally shown on a p_{me}–n diagram, i.e., mean effective pressure versus engine speed. The working fluid air with constant properties those at the inlet, i.e., cold, the cold-air standard model focuses on that makes analysis easy.

The energy exchanges for the trapped control mass, m, are

$W_{12}/m=cv(T_2-T_1)$, $Q_{12}=0$, $W_{23}=0$, $Q_{23}/m=c_v(T_3-T_2)$, $W_{34}/m=c_v(T_4-T_3)$, $Q_{34}=0$, $W_{41}=0$, $Q_{41}/m=c_v(T_1-T_4)$, and the energy and exergy efficiencies are:

$$\eta_{e.Otto}=\frac{W_{34}-W_{12}}{Q_{23}}=\frac{(T_3-T_4)-(T_2-T_1)}{T_3-T_2}=1-\frac{T_1}{T_2}=1-\frac{1}{r^{\gamma-1}}\text{ and }\eta_x=\frac{W}{W_{max}}=1-\frac{1-\frac{T_1}{T_2}}{1-\frac{T_1}{T_3}}$$

The working fluid is not recycled by the real spark-ignition engines; they are all internal combustion engines in which ambient air is pumped in and fuel that is the volatile liquid such as gasoline or gas is added to prepare a homogeneous reactive mixture that is ignited by an electrical spark at a certain point cinematically or electronically controlled.

A minute spark is sufficient to ignite the easily flammable fuel or mixture of air. The addition of fuel is now electronically controlled by injection, it can be at the inlet of each cylinder or directly inside the cylinder, as in a modern diesel engine (but with a much lower injection pressure of less than 10 MPa); at first, perhaps from 1900 to 1980, the carburetor was the standard fuel-addition system, based on venturi suction from a small float-regulated fuel chamber.

Gaseous fuels, such as coal-gas, liquefied petroleum gas (LPG) and NG, are also used in place of gasoline, as are other liquid fuels, such as methanol, ethanol, and ethers, generally added up to 20% to gasoline.

With the advent of direct injection into the cylinder with electronic control, mixture stratification has become possible, highly decreasing emissions and dangers of detonation at part loads as it was already suggested by Otto himself, but was not viable at the time; the idea is to start with a lean dosage and end with a stoichiometric mixing close to the spark plug at ignition time, controlling engine power through fuel injection such as in diesels, in particular by choking the main air flow as in homogeneous-mixture conventional Otto engines.

One of the main advantages of reciprocating internal combustion engines, both Otto and Diesel type, is that cheap materials can be used in their construction (e.g., cast iron against very expensive nickel alloys in gas turbines), because peak temperatures (up to 3000 K) are achieved only within the burning gases and for short periods, with an average temperature of around 700 K throughout the cycle; which would be the quasi-steady temperature level at the wall (because of its much larger thermal inertia), even without cooling; engine-wall cooling is applied to lower this adiabatic temperature down to 600 K to prevent thermal decomposition of lubricating oil.

Reciprocating engines can be two-stroke or four-stroke engines in a cycle; the simple to model is the four-stroke engine:

- One stroke is used to fill the cylinder with the air-fuel mixture through the inlet valve (the inlet pressure is slightly below atmospheric pressure);
- The second stroke is used to compress the mixture with closed valves and to burn it with a spark at a precise point near the maximum compression; burn it with a spark; and
- A third stroke is used to let the product gasses to expand and do work on the crank-shaft; and a fourth stroke is used to force the burnt gasses out through the exhaust valve, and start a new cycle (two crank-shaft turns have elapsed).

There is a difference between intake and exhaust in two-stroke engines, with the inlet stream used to sweep the burnt gasses (throwing unburnt gases to the tailpipe), and the whole process is completed in one turn of the crankshaft. Despite the higher fuel consumption and pollutant emissions, and the problems in lubricating them and oil is applied to the fuel, two-stroke engines have very high specific power and are thus used for small engines in model cars, garden engines, bikes, and outboard boats.

The design and operation of the valve is crucial for all reciprocating engines, but especially for the two-stroke models. Spark is created by high-voltage dielectric breakdown in the spark-plug gap, nowadays with local coils excited by solid-state electronics (e.g., thyristor trigger and Hall-effect distributor), and historically (say from 1910 to 1980) with a central induction-coil, a mechanical contact breaker, and a mechanical distributor.

For homogeneous mixing engines, the upper limit in power and engine size is dictated by the burning time, i.e., the propagation time of the initial flame ball created around the spark, to the chamber corners. Actual propagation speeds are 20 to 30 m/s, proportional to the mean turbulence level (causing flame stretching); there is a fortunate connection between this flame speed and engine speed, as turbulence is produced in the intake process, and increases with flow rate, so that the combustion process extends for almost the same crank angle, about 30 degrees, for all engine regimes.

Such turbulent flame speed is well above the laminar flame speed, which for all typical fuel mixtures is below 0.5 m/s (note that only a 5 cm run in a car engine cylinder would last 0.1 s at 0.5 m/s, but at 3000 rpm its crankshaft has gone 50–0.1 = 5 turns).

Figure 9.5. *Refrigerator works on the Principle of thermodynamics.*

Sources: Image by Flickr.

9.4. AIR CONDITIONERS AND REFRIGERATORS

In a warm atmosphere air conditioners and refrigerators are built to cool down everything. As with heat pumps, heat transfer from cold to hot involves work input, and that is costly. According to how much work input is needed, the efficiency of air conditioners and refrigerators is determined by how much heat transfer takes place from a cold environment.

What is known to be the gain of a heat pump is regarded as waste heat in a refrigerator. They therefore describe the performance coefficient of an air conditioner or refrigerator to be Noted again that, as we can see, an air conditioner will have a lower performance coefficient than a heat pump because it is greater than.Real air conditioners and fridges typically perform exceptionally well, with values varying from 2 to 6. Such numbers are better than the figures for the above-mentioned heat pumps, because the variations in temperature are greater, but they are lower than those for Carnot engines working between the same two temperatures.

A form of rating system has been developed, called the energy efficiency rating. This classification is an example of non-SI units still being used and being important to customers. To make it easier for customers, Australia, Canada, New Zealand, and the United States use a5-star Energy Star Rating the more stars, the more energy-efficient the appliance is.

It is measured in British thermal unit (Btu) mixed units per hour of heating or cooling separated by the power input in watts. Room air conditioners with varying from 6 to 12 are readily available. Though not the same as the one just mentioned, these are useful for comparison purposes—the greater the, the cheaper it is to run an air conditioner (but the higher its purchase price is likely).The sum of heat transfer from a cold environment in Btus, time in hours, work input in joules, and time in seconds can be represented as an air conditioner or refrigerator.

Figure 9.6. *Air conditioners works on principles of thermodynamics.*

Sources: Image by Flickr.

9.5. SUSTAINABLE

Nowadays the word sustainability is often used in many different contexts. In the field of engineering, for example, there have been arguments regarding sustainable goods, sustainable production, sustainable designs and so on. Although sustainability represents a very serious human problem, the word seems to have been overshadowed by many of those usages.

That we give these preliminary thoughts is in the context of facilitating a serious discussion about this idea.

Sustainability statement by the Brundtland UN Commission says sustainable development is the growth that meets the needs of the present generation without undermining future generations ability to meet their own needs. The assertion poses several problems that are laden with meaning, and at first glance seems unworkable.

For example, what one person wants might be considered unnecessary consumption to another. Therefore, humans have to express and articulate their desires for future generations? What is more, what development means is of crucial importance, particularly development needs growth, and if so, what kind of Kenneth Arrow and a group of economists and ecologists have taken on the task of translating this statement into an essentially economic framework.

Basically, they reinterpret this statement in terms of maintaining a certain form of resource base known as genuine investment to ensure that the rate of change of aggregate utility based on total demand must be increased or at least equal to zero.

Genuine investment has components, Manufactured resources, including the machinery and factories needed for this purpose; Their resulting criterion is that the sum of these three capital forms has to be constant or that. This criterion's power is its being operational. Economists have developed methods to measure these forms of capital (albeit with some difficulties and controversy).

In reality, every year the World Bank publishes The Little Green Data Book, which keeps national accounts for many countries around the world in terms of real investment or a similar metric. The planet continues to behave sustainably according to this novel.

Despite depletions in different natural capital resources such as energy and mineral resources and CO_2 harm, these are offset by additions to manufacturing and human capital, resulting in an Adjusted net savings of

7.4% of GNI (gross national income) in 2005. Nevertheless, this aspect, that natural capital can be substituted for manufactured and human capital, is extremely controversial. This is the so-called weak form of sustainability and it means that boats are considered to be replacements for fish, as Herman Daly has put it.

9.5.1. Open System Example

The interactions of energy on a biosphere model's global scale will lead to significant difficulties in determining the causality of events within the complex structure of the closed system, even though strong correlations between the collection of properties and interactions could be identified.

The knowledgeable approach therefore to the concept of sustainability criteria must therefore be to:

- define a set of critical interacting open sub-systems that constitute the eco-sphere;
- recognizing the states of these sub-systems in accordance with existing interactions and their properties; and
- formulate resource flows as material and energy interactions between the sub-systems. In theory, each of these sub-systems may have well-defined end states though one may not need to be aware of all the intermediary states at the same time.

In General, the state of a global sustainability model for each sub-system would not be in a stable state. It is also possible to conclude that each sub-system would not be in equilibrium, depending on the conditions of the state and interaction.

Following the minimum set of considerations set out above, an example of a reasonably simple global eco-sphere model that interacts with techno-sphere involving humans with their production base and institutions would consist of four, energy, and material interacting, thermodynamically well-defined sub-systems.

This means that open, non-isolated sub-systems interact through energy and material interactions in the respective states, resulting in simultaneous changes in the mass, energy, and entropy of the sub-systems and also the other properties may or may not change. The rate of mass, energy, and entropy change can be interpreted as exergy changes for each subsystem.

This exergy change can be interpreted as an equivalent increase in the available energy for the interactions of sub-systems expressed in terms of thermodynamically appropriate work interactions though in the given context, it may not actually be a work interaction. The magnitude of this change in non-system exergy time rate can be used as one of the sustainability metrics of a given system.

This would be supported on the basis that the available energy reflects a given sub-system's intrinsic ability to provide a driving force in interactions required to change the state of the overall system.

If one presumes that some aspects of each sub-system's sustainability can be measured by a positive or a limit equal to zero exergy rate change, it should be clear that such change must be a consequence of both

- interactions; and
- related changes in the state of the sub-systems, i.e., ultimately depending on the efficiency of internal transformations.

It seems obvious that a single subsystem's sustainability cannot be either necessary or adequate to create a combined collection of subsystems or the overall system's sustainability. Although exergy is not conserved, it is additive, however, and certain sustainability characteristics of interacting subsystems can be measured in terms of aggregate exergy change.

It is a fact that the system as a whole interacts with the rest of the universe by energy (dominant) and material (weak) interactions.

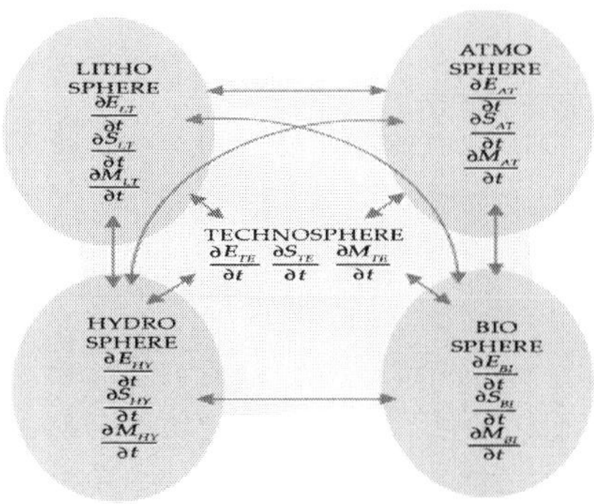

Figure 9.7. *Four subsystems of technosphere.*

Source: Image by MIT.

The Techno-sphere and four sub-systems of ecosphere interacting among themselves through energy and material interactions (arrows indicate interactions).

This system enables the determination of the exergy supply rate and the exergy utilization rate of both material and energy flows. Thus, one may evaluate the entropy or exergy loss generation of each of the sub-systems.

In the case of the technologosphere, one might contend that the loss of availability of energy can be compared by exergy conversion efficiencies to the state of the current state of development by, say, the level of technology development. For a transient, open system, one can show that the following relationship holds:

$$T_0 \dot{S}_{irr} = -\left(\frac{dE}{dT} - T_0 \frac{dS}{dt}\right) - \sum_j \dot{W}_j + \sum_k \dot{Q}_k (1 - \frac{T_0}{T_K}) +$$
$$\sum_{in} \dot{m}(h - T_0 s) - \sum_{out} \dot{m}(h - T_0 s) + \sum_{in} \left[\frac{1}{M}(\mu^* - \mu_0)\dot{m}\right]_i$$
$$-\sum_{out} \left[\frac{1}{M}(\mu^* - \mu_0)\dot{m}\right]_i$$

The equation's left-hand side reflects the exergy lost on any of subsystems. or to the whole system. Such loss reflects the combined loss or gain in availability caused by the removal of resources or the deposition of waste within the subsystems as well as all irreversible transformations within the system.

The first term on the right hand side shows the time rate of change in the availability of the considered open sub-system as a result of energy and entropy rate changes, whereas the other four terms account for interactions of exergy induced by both energy (work rate and heat rate) and material flow interactions (in and out).

The last two terms are the rates of chemical exergy generated by the corresponding mass flow rates. The example of an open sustainability sub-system analysis for sustainability can be thought as the hydrosphere and technosphere interaction. The aim would be to discover the amount of exergy one required to be invested in the form of energy resources for desalination in a way to replenish freshwater withdrawals.

The hypothetical situation entails a worldwide calamity that would prohibit the use of fresh water from both surface and land. In this scenario, therefore, the supply must be provided by desalinating the seawater from the oceans but using the lowest amount of renewable energy resources required. In order to solve this issue, an open system can use the exergy balance equation to separate pure water from the salts in the saline water and use solar energy with an optimal conversion to the exergy rate.

This process occurs between the domain of Hydrosphere and Techno sphere and includes an open sub-system that takes saline and provides fresh water. In this assumed scenario, the withdrawal rate for the year 2000 would be specified based on the withdrawal rate in the US but estimated for the entire world to 2030. In these situations, it would be useful to decide whether the available sea water as a resource could be sufficient and what the volume of renewable energy resources for the task to be completed would be.

The calculation of energy would include three cases:

- the minimum estimated energy rate needed to distinguish the amount required for global supply in 2030 at the rate of US withdrawal in 2000;
- the energy rate required for the same task as but using the most efficient, real-current technology; and
- the energy rate for the same task as in but using the technology from the end of the previous century.

The minimum use of energy resources is an energy (exergy) available which is needed to separate NaCl from H2O in saline water. By definition, this amount is equal to the destruction of exergy in that process, that is, equal to the product of entropy generation and environmental temperature.

Assuming that there are no other heat and function interactions, in a steady state, entropy balance equation reduces the desalination process equality and the difference between the open system's inlet and outlet entropy rate. As it is displayed by Gyftopoulos, segregation of salts from marine water can be regarded at any time while separating a small number of mineral components of the mixture from a large amount of marine water. This results in the determination of entropy change in the process which can represented as follows:

$$\Delta s = -((R / M_{H_2 0}) \ln (y_{H_2 0})) = -((R / M_{H_2 0}) \ln (1 - y_{NaCl})$$
$$\sim ((R / M_{H_2 0}) y_{NaCl})$$

where y's represent mol fractions of the constituents (it is assumed that saline water has 35,000 ppm of NaCl salt in it). The change in entropy rate, under the conditions defined, is equal to the entropy generation rate. The minimum exergy rate to be supplied to the separation process can be determined by multiplying the entropy generation rate with surrounding temperature that is assumed to be 25°C.

The data used are The U.S. freshwater withdrawal was 1.3*1012 kg per day in 2000. Taking into consideration that the US population in 2000 was

285M and the same per capita withdrawal for the world population in 2030, the required annual withdrawal would be 1.5*1016 kg per year.

The minimum exergy expenditure (a result of entropy generation and surrounding temperature) would be 1.4096 kJ/kg based on the mentioned open system model. So, the exergy rate usage in 2030 would be approximately 1 TW for this task alone. Indeed, this is a significant amount of energy.

More precisely, the estimated consumption of each type of energy resource that is oil, gas, coal, nuclear, and other in the year 2030 would be of the same magnitude, that is, between 1 and 8 TW. One should also bear in mind that the situations thought here provides the freshwater withdrawal rate of energy consumption based on an ideal, thermodynamically minimum exergy rate for the process of separation.This is not the technical limit for a given task at a given time and level of technological development but the minimum, theoretically possible number, no matter how high the level of technology may be. From the sustainability point of view, it might be interesting to know from the viewpoint of sustainability that the most advanced desalination processes that is presently under development require an exergy use of at least four times greater.The present standard technology would need roughly eight times more than the most efficient state of the art system. Lastly, one may think that mid-last century technology would require at least an order of magnitude of higher exergy expenditure. The crucial step in implementing this resource management strategy and determination of resource loss availability is to identify a state of each sub-system.

This step is important in practical application of the method and must include hypotheses regulated by the degree of rigor imposed. Finally, it is emphasized that the proposed mechanism for the sustainability of a sub-system and or overall eco-sphere system in terms of thermodynamics, i.e., exergy rate change or entropy generation in terms of energy units and the result between the temperature factor and entropy generation is not sufficient, but it is a required criterion for sustainability.

In other words, it is not suggested that a single physical criterion can be used to determine ecosphere sustainability that would include the techno-sphere of dynamic non-physical interactions. It is stated the physical criterion based on thermodynamics like the entropy generation and exergy destruction cannot either be regarded as a single criterion, just as one cannot assume a weak sustainability criterion based on the time rate of change of genuine investment in the economic framework to be a single criterion. There are multiple criteria that can be used.

9.6. THERMODYNAMIC TECHNIQUES USED IN FOOD CHEMISTRY

The varied biophysical techniques are used in the processing and preservation of food materials and in the manufacture of value-added products, along with thermal and non-thermal methods. The processing of foods aims to inactivate spoilage and pathogenic microorganisms during storage. Alteration occurs in food products during processing including carbohydrates, lipids, vitamins, and, most notably, proteins.

These changes result in structural and functional changes in micro- and macromolecular foods affecting the physical, organoleptic, and nutritional properties of the food. The Food materials are complex biological systems with a heterogeneous, heterophase mixture of components of high and low molecular weight and their aggregates and complexes. The food structure may include a wide range of three material states including dilution to concentrated liquids, solids, and mixtures of multi-liquid, liquid-solid, liquid-gas, and solid-gas structures.

Many biophysical techniques are used to address the wide range of structures and compositions that develop a basic understanding of processing and storage conditions impact. The data resulting from these types of research can be used to predict the physical properties of foods to improve the conditions for food processing and storage.

Calorimetry is especially appropriate for analyzing food materials. In particular, since many methods of food processing require thermal treatment (heating, cooling, freezing) of materials, thermal characterization of food systems and their components contributes to data that can be directly related to processing protocols.

Determining the thermal properties of food materials, such as specific heat as a function of temperature, is essential in calculating heat transfer and energy balance. The generation of a credible database to create equations predicting the thermal properties of food materials can be accomplished by using calorimetry to optimize food processes.

Food materials and their components also undergo conformation transitions and phase transitions. Lots of physicochemical effects can be found in the temperature range between -50°C and 300°C while analyzing foods using thermal analysis and calorimetric techniques.

The calorimetry data can be analyzed to assess the thermal and thermodynamic stability of different phases for a rational design of formulations and process conditions for food products. Calorimetry can be used to measure for the response to some perturbation caused by heat effects.

The use of thermodynamic techniques in food science is provided by two distinct ultrasensitive calorimetric techniques namely Isothermal Titration Calorimetry (ITC) and Differential Scanning Calorimetry (DSC).

In DSC disturbance is a change in sample temperature whereas in ITC the disturbance is the introduction of new material into the sample. Even though the calorimetric technique is powerful, the validity and usefulness of the data are strongly dependent upon careful use of the equipment and correct data interpretation.

Calorimetric data are strongly dependent on the conditions maintained during the experiment because of the ultrasensitive instrumental setup. At the time of selecting the calorimetry parameters, one should follow the guidelines provided by Haines-

- **Time Scale:** In dynamic measurement systems in particular, the experimental time scale will match the time scale of the observed event for events to be detected.
- **Heat Flow Magnitude:** If the energy associated with the change is high, it can cause ambiguities in its detection. Increasing the scanning rate increases the signal; however, it can cause deviation from the conditions of equilibrium, which requires models beyond the standard thermodynamics treatment of calorimetric data.
- **Loss of Moisture during Experimentation:** Biological samples are usually products with a high moisture content. If the sample cell is not sealed properly, the sample's moisture content can change over the course of the experiment due to evaporation. This can result in an overestimation of the transition temperature as well as a decrease in the transition enthalpy.
- **Interpretation of Overlapping Peaks:** Biological samples can contain multiple components at similar temperatures which undergo thermally induced transitions.

The advantages of using calorimetry for the study of food components are:

- Calorimetric methodology does not impose any preconditions on the experimental sample's physicochemical properties unlike the spectroscopic method to have in pure state or chromophore presence.
- It is feasible to model an independent and direct enthalpy change

(ΔH) model and to change the heat capacity at constant pressure (ΔHP).

- Materials must not be uniform or even a homogeneous mixture. In addition, beyond pure materials, the technique can be used to analyze the interactions between the components in a complex system, and the way the processing changes the interactions.
- The preparation of samples for calorimetric analysis is very straightforward and can be done by the researchers without any problems. In present times the instruments are highly developed to measure thermal events accurately. The theory behind the technique is well known, thus facilitating data interpretation.

9.6.1. DIFFERENTIAL SCANNING CALORIMETRY

Differential Scanning Calorimetry (DSC), that measures the heat capacity as a function of temperature, is a well-established thermal analysis method that detects and monitors thermally induced conformation transitions and phase transitions as a function of temperature.

Numerous peaks or inflection points (one to several) representing the thermally induced transitions can be observed during temperature scanning, depending on the complexity of the material. The direction of the peak corresponds to the transition nature, being either heat-absorbing (endotherms) or heat-releasing (exotherms).

While solid melting and protein denaturation exhibit endotherms, carbohydrate crystallization and protein aggregation manifest as exotherms. DSC is a thermoanalytical methodology where a temperature feature calculates the difference in the amount of heat needed to increase the temperature of a sample and reference.Throughout the experiment both the sample and the reference are held at almost the same temperature. Normally, the temperature program for a DSC analysis is generally designed such that the temperature of the sample holder increases linearly as a function of time. The reference sample should have a well-defined heat capacity over the temperature range that must be tested.The basic principle underpinning this method is that when the sample undergoes a physical transformation, such as phase transitions, it will require more (or less) heat to flow to it than the reference to maintain both at the same temperature. Whether the heat will flow to the sample more or less depends on whether the phase is exothermic or endothermic.

Differential scanning calorimeters are able to measure the amount of heat consumed or emitted during such transitions by measuring the difference in heat flow between the sample and reference. Thus, a curve of heat flux versus temperature is the product of a DSC experiment. DSC can be used to measure various characteristic properties such as solid crystallization temperature, sample oxidative stability, used in the pharmaceutical and polymer industries, etc.

Temperatures are measured using a calorimeter for the endothermic and exothermic transitions and the heat involved in such transitions. Inflection points reflect glass transitions; that is, transitions from a glassy state to rubbery state. The temperatures of the transition (Tpeak/Tg/Tm) represent the thermal stability of the transition phase or state.

In addition to evaluating the bulk heat capacity of the material, one can derive from calorimetry data values for the thermal and thermodynamic changes in free energy (ALE), enthalpy (ΔH), entropy (ΔLE), and heat capacity (ΔLE) of the various transitions.

In this method the heat flow to the sample and reference remains the same instead of the temperature. After equal heating of the sample and reference, variations in phase and other thermal processes create a temperature difference between the sample and reference.

Similar information is provided in both DSC and DTA (differential thermal analysis). DSC measures the energy needed to keep both the reference and the sample at the same temperature whereas DTA measures the temperature difference between the sample and the reference when both are placed under the same heat. DTA is an older technique compared with DSC.

So, DSC is better than DTA and more sophisticated and improved. DTA device can be used in very high temperatures and in aggressive environments where DSC device may not work. In DSC the effect of sample properties on the peak region is comparatively less than in DTA. The basis for thermodynamic analysis of food materials is that it is possible to define the related initial and final states (pre-processing and post-processing states) and to calculate the energetic and structural differences between these states using calorimetric instruments.

The calorimetry can be used to evaluate the effect of other physical and chemical variables by comparing the thermo grams of the materials before and after exposure to the variable outside the calorimetry. However, in most food processing food ingredients are mixed or diluted with a liquid (water, milk) or with a powder (sugar, salt, and yeast). For simulation of such transformations and interactions, the limited volume and the lack of in mixing constitute the major drawbacks of the DSC technique.

9.4.2. Isothermal Titration Calorimetry

One or more binding event is needed by most of the biological process. There are various types of binding reactions including, but not limited to, assembly of protein subunits in functional enzyme complexes, the development of enzyme-inhibitor complexes, the creation of proteinnucleic acid complexes, enzyme-substrate binding, and enzymecofactor binding.

These binding processes may be described with standard thermodynamic parameters in mind. Measurement of the binding thermodynamic parameters will achieve a predictive understanding of the binding process. All of the above-mentioned binding processes can be studied by some variant of the ITC experiment.Because the binding sites for small molecules to proteins appear to be well-defined and low in number, and because most binding reactions require a non-zero enthalpy shift, protein-small molecular interactions are often particularly suitable for isothermal titration calorimetry (ITC) analysis.

Here, one can thought a simple association without reaction described by the equilibrium $nL + M \leftrightarrow MLn$ of a small molecule L, with a protein or other macromolecule M, with an identical, non-interacting binding sites. A calorimetric isothermal titration device consists of two identical cells consisting of a highly efficient thermal conducting material (Hasteloy) surrounded by an adiabatic shell.The sensitive thermopile or thermocouple circuits sense changes in temperature between the two cells and the cells and the jacket. Where necessary, heaters located on both cells and the jacket are activated to maintain identical temperatures between all components. The macromolecule solution is generally placed inside the sample cell in an ITC experiment.The reference cell comprises of buffer or water except that of the macromolecule. A constant power (< 1 mW) is added to the reference cell before titrant is injected. This signal commands the feedback circuit to activate the sample cell heater located on it. That reflects the signal from the base line. The direct measurable calculated in an ITC experiment is the time-dependent power input needed in the sample and reference cell to maintain equal temperatures.

Heat is taken up or produced during the injection of the titrant into the sample cell depending on whether the macromolecular interaction reaction is endothermal or exothermic. In an exothermic reaction, the temperature in the sample cell will grow and the feedback control will be deactivated to retain equivalent temperatures in endothermic reactions between the two cells, the reverse will occur indicating that the feedback circuit must increase the sample cell's power to maintain the temperature.During a calorimetric titration the heat absorbed or evolved is proportional to the fraction of the bound ligand. All or most of the added ligand is bound to the macromolecule for the initial injections, resulting in broad exothermic or endothermic signals depending on the type of the interaction. With the increase in concentration of the ligand, the macromolecules become saturated and consequently less heat is produced or absorbed by additional titrant.The quantity of heat produced by adding ligand may be expressed by the equation $Q = V_0 \Delta H_b [M]_t K_a [L] / (1 + K_a [L])$ in which Vo is the volume of the cell, where ΔHb is the enthalpy of binding per mol of ligand, where (M) t is the total concentration of macromolecule comprising bound and free fractions, where Ka is the binding affinity and (L) is the free concentration of ligand.

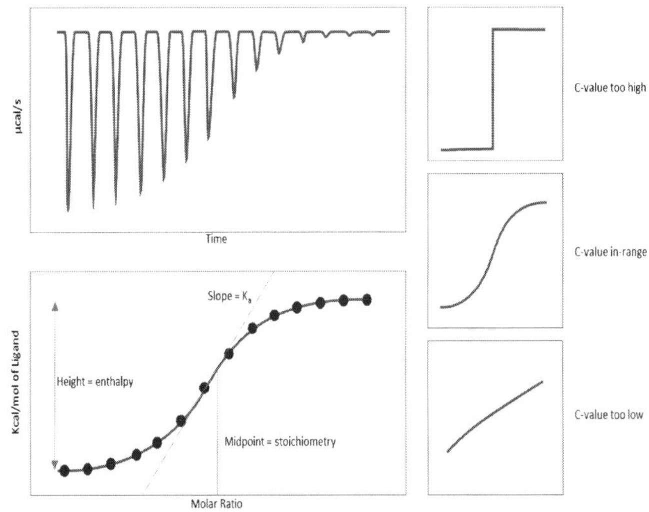

Figure 9.8. *Isothermal titration calorimetry.*

Sources: Image by Wikimedia.

9.7. CONCLUSION

The thermodynamics has played a great role in advancement of human civilization since the beginning of industrial revolution. The rapid advancement of automobiles to modern cars and from steam engines to heavy duty pumps al work on the principles of thermodynamics. The pumps used in nuclear power plants to cool the reactors also work on this principle. The modern electronic machines like refrigerators and air conditioners also work on the principle of thermodynamics. The thermodynamics is also applied in the food chemistry.

The application of thermodynamics in the machines and devices that the human beings use is huge in day to day life.

REFERENCES

1. Churala, P. P., (n.d.). *An Overview on the Thermodynamic Techniques Used in Food Chemistry*. [online] Available at: http://dx.doi.org/10.4172/2329–6798.1000142 (accessed on 11 March 2020).
2. Gutowski, T., Sekulic, D., & Bakshi, B., (2009). *Preliminary Thoughts on the Application of Thermodynamics to the Development of Sustainability Criteria*. [ebook] p. 2. Available at: http://web.mit.edu/ebm/www/Publications/IEEE-May-09.pdf (accessed on 11 March 2020).
3. Khemani, H., (n.d.). *Applications of Second Law of Thermodynamics*. [online] Bright Hub Engineering. Available at: https://www.brighthubengineering.com/thermodynamics/4349-applications-of-second-law-of-thermodynamics-with-automobile-engines/ (accessed on 11 March 2020).
4. Opentextbc.ca. (n.d.). *Applications of Thermodynamics: Heat Pumps and Refrigerators – College Physics*. [online] Available at: https://opentextbc.ca/physicstestbook2/chapter/applications-of-thermodynamics-heat-pumps-and-refrigerators/ (accessed on 11 March 2020).
5. Thermal Machines and Heat Engines, (n.d.). [ebook] p. 3. Available at: http://webserver.dmt.upm.es/~isidoro/bk3/c17/Power.pdf (accessed on 11 March 2020).

INDEX

N

O

P

Q

R

V

W

Z